TECHNOLOGY, A STUDY OF MECHANICAL ARTS AND APPLIED SCIENCES

ISBN: 978-1-291-58550-6

I0397390

Andreas Sofroniou   2013 ©   Copyright

Andreas Sofroniou   2013 ©   Copyright

TECHNOLOGY, A STUDY OF MECHANICAL ARTS AND APPLIED SCIENCES

ISBN: 978-1-291-58550-6

1

CONTENTS   PAGE:

Andreas Sofroniou

3

Andreas Sofroniou

1. FUNDAMENTAL TECHNOLOGIES

Technology, the study of the mechanical arts and applied sciences, although the precise meaning of technology has changed over the years and is still to some extent fluid. Many fundamental technologies -- the smelting and working of metals, spinning and weaving of textiles, and the firing of clay, for example -- were empirically developed at the dawn of civilization, long before any concept of science existed.

With the advent in about 3000 BC of the first major civilizations in Egypt and Mesopotamia (and a little later in India and China), many new technologies were developed; irrigation systems, road networks and wheeled vehicles, a pictographic form of writing, and new building techniques.

Other civilizations subsequently became important technological centres, notably those of Greece and Rome, the Arab empire of the 7th to 10th centuries, and the Mayan, Aztec, and Toltec civilizations of meso-America.

In the mid-16th century the focus of technological change shifted to Europe, with the beginning of the Scientific Revolution. This was both an intellectual revolution and a practical one, questioning established dogma, reinterpreting old ideas, and seeking to advance knowledge of the natural world by observation and by experiment.

Initially the new ideas and techniques engendered official persecution, but by the mid-17th century the tide of opinion had changed, as indicated by the formation under royal patronage of the Academie des Sciences in France and the Royal Society in Britain.

By the late 17th century, technology essentially meant engineering, as is indicated by the title of a British book by T. Phillips, published in 1706: *Technology: A Description of Arts, Especially the Mechanical.*

Half a century later, however, Diderot's monumental twenty-eight volume *Encyclopedie* (1751-72) encompassed not only the mechanical but also what he called the liberal arts, including glass-making, agriculture, brewing, and soap-boiling.

In the UK in 1866, the teacher and author Charles Tomlinson published his three-volume *Cyclopaedia of Useful Arts, Mechanical and Chemical, Manufactures, Mining, and Engineering*, which, by including the old empirical processes as well as those that had arisen through the application of scientific knowledge, approached the modern concept of technology.

6

During the 19th century science began to create many new technologies, such as the electric telegraph, the telephone, electricity generation and supply, and photography. The trend continued into the 20th century with the introduction of many goods and services made possible only because of further advances in science.

These have included radio and television, sound recording and reproduction, synthetic fibres, a wide range of pharmaceutical products, nuclear power, and perhaps most important of all, the development of the computer, information technology, as a new technological revolution.

Since the 1970s pollution, depletion of energy resources, and other adverse effects of technology have caused increasing public concern. This has led to the growth of alternative technologies, with an emphasis on renewable energy sources such as solar and wind power, the recycling of raw materials, and the conservation of energy.

Outside the West, only the most basic technology is available to hundreds of millions of people. Tropical agriculture remains resistant to the application of science, and medical technology has made only limited impact in the Third World: according to a recent estimate by the World Health Organization, four-fifths of the world's population still have no regular access to health services of any kind.

For people still locked into subsistence agriculture, the convergence of technology and applied science, which the Western world takes for granted, is largely irrelevant. However, in recent years Western aid has sought to develop appropriate technologies, using local materials and techniques, in partnership with the indigenous peoples.

7

2. PREHISTORIC TECHNOLOGY

2.1 BEFORE WRITTEN HISTORY

Prehistoric technology refers to the technological developments occurring before written history. Although a somewhat artificial concept, in that it presupposes that technology advances uniformly world-wide, prehistoric technology generally includes the skills that were practised before the rise of the earliest civilizations in the Middle East and the appearance of cuneiform writing (about the middle of the 4th millennium BC).

This is roughly coincidental with the first known use of copper and bronze, so the working of these metals can be recognized as a prehistoric technology. Iron did not come into use until about 1200 BC; however, many different cultures in Europe used iron long before they had any mastery of written language.

Before the availability of metals, the main building materials were wood and stone. Stone was also used for axes and other tools, flint being skilfully worked to give a cutting edge to knives and arrow heads.

Leather from the hides of slaughtered animals was plentiful, and provided material not only for clothing but also for screens to give protection from the weather and for making buckets and other containers.

Basket-work and weaving originally used fibres collected in the wild, and date from around 5000 BC. The most important feature of prehistoric development was the transition from a hunter-gatherer culture to a settled way of life associated with the beginnings of agriculture, in about 9000 BC. The earliest permanent settlement for which much detailed information has come to light is Jericho, in the Jordan Valley, where the original walled city dates from about 8000 BC.

Technological innovations continued through the centuries with various other forms of technology, such as the processes of smelting and spinning.

Smelting is the process by which an ore concentrate is fused (melted) at high temperatures to extract a matte or impure metal. Usually a flux is added to remove impurities as a slag. In the smelting of iron ore, coke is added to the blast-furnace to reduce the oxide to iron metal.

8

2.2 STONE AGE TECHNOLOGY

The identification of the history of technology with the history of manlike species does not help in fixing a precise point for its origin, because the estimates of pre-historians and anthropologists concerning the emergence of human species vary so widely.

Animals occasionally use natural tools such as sticks or stones, and the creature that became man doubtless did the same for hundreds of millennia before the first giant step of fashioning his own tools.

Even then it was an interminable time before he put such tool-making on a regular basis, and still more aeons passed as he arrived at the successive stages of standardizing his simple stone choppers and pounders and of manufacturing them--that is, providing sites and assigning specialists to the work.

A degree of specialization in tool-making was achieved by the time of Neanderthal man (70,000 BC); more advanced tools, requiring assemblage of head and haft, were produced by Cro-Magnon *Homo sapiens* (perhaps as early as 35,000 BC), while the application of mechanical principles was achieved by pottery-making Neolithic man (6000 BC) and by Metal Age man (about 3000 BC).

2.3 TECHNOLOGICAL UNITY

The material that gives its name and a technological unity to these periods of prehistory is stone. Though it may be assumed that primitive man used other materials such as wood, bone, fur, leaves, and grasses before he mastered the use of stone, apart from bone antlers, presumably used as picks in flint mines and elsewhere, and other fragments of bone implements, none of these has survived.

The stone tools of early man, on the other hand, have survived in surprising abundance and over the many millennia of prehistory important advances in technique were made in the use of stone. Stones became tools only when they were shaped deliberately for specific purposes, and, for this to be done efficiently, suitable hard and fine-grained stones had to be found and means devised for shaping them and particularly for putting a cutting edge on them.

Flint became a very popular stone for this purpose, although fine sandstones and certain volcanic rocks were also widely used. There is much Palaeolithic evidence of skill in flaking and polishing stones to make scraping and cutting tools. These early tools were held in the hand, but gradually ways of protecting the hand from sharp edges on the stone, at first by wrapping one end in fur or grass or setting it in a

9

wooden handle, were devised. Much later, the technique of fixing the stone head to a haft converted these hand tools into more versatile tools and weapons.

With the widening mastery of the material world in the Neolithic Period, other substances were brought into the service of man, such as clay for pottery and brick; and increasing competence in handling textile raw materials led to the creation of the first woven fabrics to take the place of animal skins. About the same time, curiosity about the behaviour of metallic oxides in the presence of fire promoted one of the most significant technological innovations of all time and marked the succession from the Stone Age to the Metal Age.

2.4 TOOLS AND WEAPONS

The basic tools of prehistoric peoples were determined by the materials at their disposal. But once they had acquired the techniques of working stone, they were resourceful in devising tools and weapons with points and barbs.

Thus the stone-headed spear, the harpoon, and the arrow all came into widespread use. The spear was given increased impetus by the spear-thrower, a notched pole that gave a sling effect. The bow and arrow were an even more effective combination, the use of which is clearly demonstrated in the earliest "documentary" evidence in the history of technology, the cave paintings of southern France and northern Spain, which depict the bow being used in hunting.

The ingenuity of these primitive hunters is shown also in their slings, throwing-sticks (the boomerang of the Australian Aborigines is a remarkable surviving example), blowguns, bird snares, fish and animal traps, and nets. These tools did not evolve uniformly, as each primitive community developed only those instruments that were most suitable for its own specialized purposes, but all were in use by the end of the Stone Age.

In addition, the Neolithic Revolution had contributed some important new tools that were not primarily concerned with hunting. These were the first mechanical applications of rotary action in the shape of the potter's wheel, the bow drill, the pole lathe, and the wheel itself. It is not possible to be sure when these significant devices were invented, but their presence in the early urban civilizations suggests some continuity with the Late Neolithic Period.

The potter's wheel, driven by kicks from the operator, and the wheels of early vehicles both gave continuous rotary movement in one direction. The drill and the lathe, on the other hand, were derived from the bow and had the effect of spinning the drill piece or the work-piece first in one direction and then in the other.

Developments in food production brought further refinements in tools. The processes of food production in Palaeolithic times were simple, consisting of gathering, hunting, and fishing. If these methods proved inadequate to sustain a community, it moved to better hunting grounds or perished. With the onset of the Neolithic Revolution, new food-producing skills were devised to serve the needs of agriculture and animal husbandry.

Digging sticks and the first crude ploughs, stone sickles, querns that ground grain by friction between two stones and, most complicated of all, irrigation techniques for keeping the ground watered and fertile-- all these became well established in the great subtropical river valleys of Egypt and Mesopotamia in the millennia before 3000 BC.

2.5 PRIMITIVE CULTURE

in the lexicon of early anthropologists, any of numerous societies usually characterized by lack of a written language, relative isolation, small population, relatively simple social institutions and technology, and a generally slow rate of socio-cultural change. History and myth in such cultures are passed on through an oral tradition and may be the province of a person or group especially trained for the purpose.

Increasingly in the 20th century, the use of the term primitive in connection with human societies and the institutions and products thereof has been regarded as a vestige of the colonial spirit in which the discipline of anthropology was born. The term primitive, it has been charged, along with the loose equivalents savage, preliterate, and others, implies that the cultures so characterized are at an earlier stage of development than "higher," or literate, civilizations.

Contemporary anthropologists generally prefer to avoid this assumption, which they regard as simplistic. In addition, early writers frequently used these terms with the implication that such peoples were mentally and morally inferior. The term non-literate has been adopted by some scholars in an attempt to avoid negative value connotations. The limitations inherent in one culture's description of another, however, render all such designations problematic. The criteria for an objective anthropological vocabulary remain a matter of considerable debate.

11

2.6 LATE PALEOLITHIC TOOLMAKING

The fourth phase of Palaeolithic tool-making was introduced perhaps 40,000 years ago by the Aurignacian industry, a forerunner of the last and most brilliant achievements of the Old Stone Age. Extraordinary inventiveness was characteristic of the Aurignacian tradition and its several short-term successors. They can be lumped into a unit of development that spans the next 25,000 years.

Fully modern man--whose first representative is the Cro-Magnon-- emerged within this period, perhaps 35,000 years ago, during the time of the development and elaboration of rock technology, which, by providing a variety of specialized tools, mostly of the flake and blade types, at last brought materials other than rock into extensive use. It was also a time when the great plains in northern and eastern Europe carried such a heavy reindeer population, in addition to wild horses and mammoths, that it has been called the Reindeer Age.

This produced a hunting economy providing food and great quantities of bone, horn, skin, sinews, and, while the mammoth lasted, ivory; with it grew new technologies exploiting the unique properties of materials hitherto unworkable because of their hardness. This technological diversification was made possible by new techniques and rock tools, whose specialization and complexity fit them to the fresh tasks.

The most significant tool was the burin, or graver, a stout, narrow-bladed flint able to scrape narrow grooves in bone; two parallel grooves, for example, would allow a sliver of bone to be detached as stock for a needle, pin, awl, or other small object. Larger pieces of bone were worked into hooks with one or more barbs or points. Sections of antler were carved into splitting wedges to work out long pieces of bone to form the dart-like projectiles of the spear-thrower. Sandrock polishers were added to the tool kit to sharpen and shape tips, needles, and other articles.

A spectacular item that developed by the end of the Palaeolithic was the spear-thrower, a hand-held stick, of wood or antler, notched at one end. Functioning as an extension of the arm, it added considerable kinetic energy, and therefore range, to a short spear tipped with flint or bone. The tipped projectile represented still another innovation, for it was the first hafted implement.

Hafting, or the fitting of a handle to a cutting edge, was a momentous and far-reaching invention of about 35,000 years ago. It was a critical step toward the creation of new tools and improved models of old ones. In its simplest form, the haft may have been no more than a grass or

12

leaf bundle whose limited function was to protect the hand when a fractured rock was used as a knife.

Mechanically, the handle became a force-transmitting intermediary between the source of the force and the tool-head. An extension of the arms, the handle provided an increased radius of swing. This moved the tool-head faster to give it more kinetic energy for a harder and more telling blow than the arms alone could provide. A man using a hand-held axe-head could cut only small trees, whereas with a hafted axe he could fell a tree of almost any size.

The prepared-core technique that provided pre-shaped flakes was refined and extended to provide pre-shaped blades, long, slender pieces of flint of trapezoidal cross section, each corner having a straight cutting edge without the serrations of a chipped tool. This is known as the blade-tool industry, a final complement to the core- and flake-tool technologies. Such blades made thin and splendid knives of great variety; many of these knives were backed; that is, the back of the blade was blunted for safer handling.

Thin blades were further reduced to smaller pieces, often having a geometric form such as triangular, square, or trapezoidal, called microliths. These small bits of sharp flint were cemented (using resin) into a groove in a piece of wood to form a tool with a cutting edge longer than it was feasible to produce in a single piece of brittle flint; examples are a spear with a long cutting edge or the farmer's sickle of later date.

The second major mechanical invention of the Upper Palaeolithic was the bow, a device even more effective than the spear-thrower for increasing the distance between the hunter and the hunted. It is difficult to date precisely, for the only evidence of its use is found in cave paintings. Mere finds of rock points without bows prove nothing because such tips were used on the projectiles of spear-throwers. The earliest representations of the bow come from North Africa from 30,000 to 15,000 BC.

Once the bow had been devised, it spread with astonishing rapidity, its effectiveness making it the weapon par excellence. When the bow was pulled, it stored the gradually expended energy of the archer's muscles; this energy was suddenly released to give the projectile a "muzzle velocity" far higher than that possible from a spear-thrower and of superior accuracy. It was a principal weapon through the 15th century AD and was ousted then only by gunpowder.

3. TYPES OF TECHNOLOGY

3.1 SPINNING

Spinning is the process of making a yarn or thread followed. This may involve the compaction and twisting of long polymer fibres, as for example, in silk processing. Alternatively, it may be by the alignment and twisting together of short (staple) fibres, such as cotton or wool.

The actual fact is that spinning of staple fibres into yarn is a very ancient craft: in India, for example, cotton was spun from at least 3000 BC. Essentially it consists of disentangling and aligning the fibres (carding), drawing them out to provide an assembly to give a yarn of the required thickness, and then inserting twist to form a coherent structure.

Early manual methods included the use of hand-held teasel cards and the distaff and spindle. Teasel cards are flat wooden plates with dried, spiky teasel plant heads attached, or rows of wire spikes, used to card (comb) the fibres. The distaff is a cleft staff on which a carded sliver of wool, cotton, or other fibre is loosely wound; the spindle is a slender, cylindrical spinning implement that is set in motion by hand, then allowed to fall, drawing out and twisting the fibres from the distaff into yarn.

The spindle and distaff were used in Europe until the Middle Ages, and are still widely used in Third World countries. The spinning-wheel was invented in India and introduced into Europe in the 14th century. It was faster than hand-spinning, and gave a more uniform yarn.

In the late 18th century the Industrial Revolution brought improved looms, which required more rapid yarn production. This promoted a spate of inventions relating to spinning: Hargreaves's spinning-jenny, Arkwright's water-frame, Crompton's mule spinning frame, and efficient mechanical carding machines to provide fibre for the spinning frames. By the beginning of the 19th century these developments had been combined into a low-cost, mechanized system for staple yarn manufacture that is essentially the same today.

The modern spinning process typically consists of the following stages. Bales of tightly packed fibre are opened, then disentangled and aligned by carding. The assemblies of parallel fibres are then drawn out by successive pairs of rollers, each pair running faster than the last: this

14

gives successively finer strands. (Wool is not drawn out in this way before twisting.)

Finally, the fibres are twisted to give cohesion, compactness, and strength. In the modern industry, a few mule frames are still in use (in the mule, the drawing out of strands and their twisting take place alternately). However, most staple yarn is now spun on ring frames or open-end spinning machines.

Silk is produced by the silkworm, which extrudes a viscous liquid through two small orifices (spinnerets) on its lower lip. The liquid solidifies on contact with air to form a pair of solid, adhering, fine protein filaments. The silk is then processed to make it into yarn.

Manufactured filaments are similarly extruded or 'spun' by pumping a long-chain polymer, in solution or in molten form, through the fine holes of a spinneret. In the 'wet spinning' of, for example, rayon and some acrylic fibres, the polymer is in aqueous solution and solidification is achieved by chemical reaction (coagulation).

In 'dry spinning' (for example, of acetate fibre and other acrylics), the solvent is organic, and solidification is through evaporation of the solvent. In the 'melt spinning' of nylon (polyamide) and other thermoplastics, solidification is by cooling of the molten polymer.

After extrusion, the filaments are pulled away from the spinneret and stretched to orient the polymer molecules parallel to the fibre axis. On some modern machines, production of filament yarn may be up to 10 km (6 miles) of yarn per minute. The continuous filaments may then be used as they are, crimped to give them more bulk and texture, or broken into short lengths for use as staple fibres

3.2 IRRIGATION

The supply of water to land to grow crops or to increase crop yields. The amount of water used or lost to the atmosphere by crops, minus the annual rainfall, determines the quantity of irrigation water needed.

In traditional irrigation systems, water is spread over the ground surface. Efficiency is between 40 and 75 per cent, and such systems are relatively inexpensive to build and maintain. Sprinkler and trickle-feed systems are generally between 60 and 80 per cent efficient, but have high capital and maintenance costs; they are suited to high-value crops.

In 'basin' irrigation--commonly used in paddy-fields for growing rice-- the ground is levelled and a low bund (embankment) retains the

15

irrigation water. 'Furrow' and 'border' methods, with sloping ground for drainage, suit crops that cannot tolerate water-logging.

All irrigation water contains some salts in solution; an increase in salinity can cause a reduction in soil fertility if the water applied is insufficient to leach the salts from around the plant roots.

3.3 ROAD BUILDING

Road is a way, often with a prepared surface, by which people, animals, or vehicles may travel between places. A street is a road through a built-up area.

The first road-builders were probably the Mesopotamians, and the earliest long-distance route was probably the 3,000-km (1,865-mile) Royal Road from Susa to Smyrna, which is thought to have been in use as early as 3500 BC. China had a fully developed road system by the end of the Shang dynasty (11th century BC); bridges were built across rivers and mountains were traversed by stone-paved stairways. By the 4th century BC the Mauryan civilization in India had also developed an extensive road network.

In Europe the pre-Roman roadways were mostly tracks worn by the feet of people and animals, or by the wheels of wagons. The first important European road-makers were the Romans (Roman roads).

They originally built roads to speed troop movements. Their roads consisted of a foundation of stone and concrete, overlaid with a camber of lime, chalk, and gravel, and surfaced with polygonal flagstones; on minor roads, a gravel or broken stone layer replaced the flagstones. Streets were laid with large polygonal blocks.

By AD 200 the road networks of the Roman, Mauryan, and Chinese empires provided trade routes from most of Europe to India and the Far East. Between the 3rd and the 18th centuries there was a road-building decline throughout Europe, the Middle East, and Asia.

In contrast, in South America this period saw the development of the Inca road system, which by the time of the arrival of the Spaniards in the early 16th century served an area of some 2 million km^2 (750,000 square miles), populated by around 10 million people. Particularly impressive were the roads running through the Andes, which included galleries cut through solid rock and extremely high retaining walls, built to support the road surface.

The modern road-building era began in 1747 with the establishment of the Ecole Nationale des Ponts et Chaussees in Paris to train engineers.

Tresaguet improved the Ecole's methods of construction, which were based upon Roman techniques, using about half the quantity of material. Before the Revolution (1789), France led the world with a network of 40,000 km (25,000 miles) of paved roads, extended later by Napoleon.

In the UK Telford built roads with a more level subsurface formation and thinner surface. His contemporary, McAdam dispensed with Telford's foundation of large stones, which was expensive. He emphasized the need to keep the natural formation dry by covering it with an impervious surface with good drainage. His road surface, known as water-bound macadam, was almost universally used until the invention of the wide inflatable tyre, which tended to disrupt the surface.

The origin of modern Tarmac(adam) dates from the 1830s, when tar was first used in Nottinghamshire, UK, to bind the surface stones. In France from 1832 onwards, powdered rock, asphalt, and bitumen mixtures were used for roads.

During the 19th century, busy urban streets were paved with hard-wearing stone blocks (often of granite) on a broken stone foundation. Concrete was first used for roads in the 1850s in Austria. By the 1920s, highways designed specifically for heavy and fast traffic were being built in the USA, and as three-lane, limited-access autostradas in Italy.

The most advanced modern road system of its time was the German autobahn network built in the 1930s. In the UK during the same period main roads, sometimes with dual carriageways, were built, but it was not until the late 1950s that the building of the motorway network began.

Motorways have at least two, usually three, carriageways. Motorways have gradual bends and gradients, designed for fast-moving traffic. Junctions are designed as roundabouts, and to avoid interrupting traffic flow, vehicles enter the motorway by means of acceleration lanes.

A feature of post-war urban road development, especially in Japan, has been the double-or even triple-decked expressways over existing streets.

Road building is now eased by the use of the bulldozer, scraper, grader, and road roller. Slip-form pavers, machines that can rapidly lay concrete on to a prepared base, are used in the building of concrete roads. In Europe, the building of major new roads in the 1990s has often attracted fierce opposition on environmental grounds

3.4 BUILDING TECHNIQUES

Archaeological evidence suggests that building as a specialized activity probably began in Mesopotamia (*c*.7000 BC), where there is little building stone. Mud buildings were generally roofed either with wooden beams supporting clay-daubed matting or with a vault (a three-dimensional arched structure).

True arches were used for openings, and methods of building vaults without centring (using a wooden framework to support the vault while building) were also known.

The Ur-Nammu ziggurat at Ur, Mesopotamia (*c*.2100 BC), had an adobe core that was faced with fired bricks bedded in bitumen and reinforced with reed mats.

Building with stone seems to have begun in the Nile valley, with the construction of important religious buildings. Although the arch was known to the Egyptians, they preferred massive columns supporting short lintels for their temples.

The Great Pyramid (*c*.2500 BC) was built with stones up to 200 tonnes each, faced with finely jointed limestone bedded in lime. Before stone-quarrying began, Egyptian buildings were made from reed bundles covered with reed matting, a technique still in use in southern Iraq.

Buildings were often built from wood where it was plentiful. Early buildings used wooden posts, planking, or entire logs to form walls and roof; later, timber frame construction was introduced to save on wood. Roofs were covered with thatching (vegetable material such as reeds or brushwood laid over the rafters) or tiles.

In Greece, this frame construction, translated into marble, is evident in such temples as the Parthenon, Athens (*c*.450 BC).

The Romans developed a strong, durable form of concrete by adding silica-rich crushed brick or pozzolana (volcanic ash) to lime. Large public buildings were roofed by concrete and masonry semicircular vaults or domes. The Pantheon dome in Rome (built in 27 BC) spans 43 m (141 feet)--a span unequalled until the 19th century.

Meanwhile, in China, there was a tradition of timber construction very different from that in the West. This type of design spread to Korea and Japan and lasted until modern times.

In Europe, the medieval mason extended Roman methods to develop slender Gothic churches with large tracery windows. Timber trussed roofs protected the masonry from the weather. Humbler buildings were still of timber, with the upper floors and roof supported by walls of closely spaced posts.

During the Renaissance, masonry remained the material for large structures, with arches being used to span openings. The domes of St Peter's, Rome (1546-64), and of St Paul's, London (1685-1710), were not buttressed externally but were encircled by iron chains to contain the outward thrust on the supporting walls below.

Increasingly, metal reinforcements were used to strengthen masonry. In France in the 1670s, Claude Perrault designed lintels for a colonnade at the Louvre, which were made from small stone blocks reinforced with an iron cage--a precursor of reinforced concrete.

By the late 1700s in Britain a number of disastrous fires in mill buildings prompted the replacement of wood with cast-iron columns and beams as the internal support for floors (external walls were still of masonry) and the development of iron windows and brick and iron vaulted floors.

In a series of large botanical glasshouses, sheet glass (which could now be manufactured in large sizes) enclosed an iron framework; masonry walls were not used at all. This technique culminated in the Crystal Palace, London (1851)--the first large building to be assembled from prefabricated components--and in the roofs of many large train stations and sheds.

The first fully framed building was probably the four-storey Menier chocolate factory built over the River Marne (to utilize water power) at Noisel, France (1871). In Chicago, development of a safe passenger lift led to the construction of skyscrapers. The Forth Bridge (1882-9) was the first major steel structure to be built in the UK.

Meanwhile, concrete had undergone significant development. In the 1760s, Smeaton had discovered that mixtures of clay and limestone produced a hydraulic cement which hardened under water--essential for his work in rebuilding the Eddystone Lighthouse off Plymouth, UK.

Subsequently, the compressive strength and reliability of these cements improved, but the concrete made with them lacked tensile strength, essential to resist the stresses in a floor or beam.

In the UK and France reinforcement techniques were developed for concrete. In the UK in 1854, A. B. Wilkinson patented beams reinforced with wire ropes and iron bars and built a reinforced concrete house (1865).

19

In France, Joseph Monier made reinforced concrete tubs for orange trees (1849), and in 1892 a reinforced concrete building frame was patented which had many features still in use today.

Modern skyscrapers have hull-and-core structures, with a light-weight curtain wall enclosing the building. The central core is a reinforced concrete tower. While reinforced concrete is also often used as the skeleton frame for blocks of flats and similar buildings, a load-bearing wall structure is usually more appropriate. The main walls run across the building to support the floors and are buttressed by internal longitudinal walls (cross-wall construction) as in the Unites d'Habitation (1947-52) by the Swiss-born French architect Le Corbusier.

For large-area single-storey buildings, both thin concrete shells and steel space frames are widely used, and by the 1980s the availability of durable polymer fabrics made possible the construction of tent-like tension structures typified by the Hajj terminal in Saudi Arabia.

A major force for the development of building techniques has been the improvement in materials manufacture and in fabrication and jointing techniques. Another has been the better understanding of structural behaviour and the evolution of mathematical techniques of structural analysis since the 16th century

3.5 INFORMATION TECHNOLOGY

Information technology (IT) is the umbrella term used to describe the practical applications of computer systems. The term has become prevalent with the increasing use of computers such as word processors for office systems, but it also embraces the more traditional areas of data processing and information retrieval.

Key factors in the recent rapid spread of microelectronics and information technology (often known simply as 'new technology') have been the drastic reduction in electronic hardware costs during the 1970s and 1980s; the so-called 'convergence' of computing and telecommunications (sometimes denoted by the term telematics); and the emergence of various formal standards and informal agreements within the IT industry.

The mass production of microprocessors and other electronic components at low unit cost has made it possible to incorporate digital electronics into a wide range of products for commerce, industry, and the home.

20

New technology has made an enormous impact on the design and performance of consumer goods, for example, cameras, video and hi-fi equipment, and personal computers. In offices and business, new technology was introduced for text processing, using dedicated (special-purpose) word processors.

Now, however, it is usual to find general-purpose small computers and workstations being exploited for a much wider range of office applications using special-purpose application software such as financial planning packages, management support software, information storage and retrieval applications, desk-top publishing, and so on. Expert systems using artificial intelligence techniques to aid decision-making can also be run on suitable personal computers.

The combination of computing with private and public telecommunication networks has been particularly significant in retailing and banking. Bar codes on products can be read automatically at check-outs, and the information used both to print a customer's receipt and to reorder necessary stocks.

Electronic funds transfer using an 'intelligent' cash register or point-of-sale terminal can then complete the cycle of electronic information flow by automatically debiting the customer's bank or credit-card account by the appropriate amount. Sales records held by the store's central computer can be used to develop overall retailing strategies; records of the purchasing patterns of particular customers can be exploited for direct marketing of goods or services.

The IT revolution has been just as great in the manufacturing and processing industries. An early application was the numerical control of machine-tools, in which digital electronics were used to control lathes and other equipment.

Now it has become common for a wide range of industrial equipment (sensors, control systems, robots, and so on) to incorporate microelectronics and to be interconnected (networked) so that information can be gathered, communicated, and processed in order to optimize the activity.

Artificial intelligence, in which computer programs mimic certain aspects of human behaviour, is beginning to play a role in areas such as process control, as well as in mechatronic products combining aspects of electronic, mechanical, and software engineering.

Again, the convergence of telecommunications and computing has been a vital factor, allowing the computer to become fully integrated into the manufacturing or other processes, thus transcending individual applications such as computer-aided design.

The introduction of new technology into areas which have not traditionally used computers has had a great influence on working practices. Some traditional tasks--or even whole categories of tasks-- have disappeared completely and new ones have appeared. In some sectors (printing in the UK, for example), adoption of new technology has led to serious industrial conflict.

Once it became commonplace for information retrieval and data processing to be carried out by digital computers in business and industrial settings, it was also natural for such information to be transmitted digitally using a range of new services; telecommunications has therefore become an essential part of many working environments.

Small computers are often linked by means of local area networks within an office or company, so that expensive resources such as laser printers or databases can be shared; a further advantage of such networking is that in-house communication can then include electronic mail, computer conferencing, or even voice messaging, as well as the traditional telephone and memo.

Digital telecommunication links extend such facilities to the outside world, providing electronic communication between widely dispersed organizations or individuals, as well as access to external databases or information services. It is even becoming possible for certain employees to work partially or entirely from home, communicating with colleagues via a computer and modem.

In the home, information is available electronically through teletex services such as Ceefax and Oracle as well as call-up videotex such as Prestel. More recently, with a growth in the ownership of home computers, individual users have linked into the Internet.

In the future, the combination of telecommunication services (including television), home computer, and new storage media such as videodisc and CD-ROM is likely to form the core of domestic entertainment and educational applications of IT.

All the above applications of new technology are based upon a computer architecture which has remained virtually unchanged since the early days of the digital computer. Alternative structures such as neural networks are currently arousing much interest, particularly for their potential in artificial-intelligence applications, while parallel computing offers possibilities for faster processing.

In the immediate future, however, further exploitation of new technology is more likely to result from a continuing expansion of telecommunications services and economies of scale in the provision of hardware than from radical changes to computer design.

Information technology raises problems of privacy and freedom of information, as data is gathered about citizens who may not have access to records held about them.

23

4. PROFESSIONAL SOCIETIES

4.1 FRENCH SCIENTIFIC SOCIETY

Academie des Sciences, the French scientific society, one of the oldest and most prestigious in the world. Founded in 1666 by Jean-Baptiste Colbert, four years after the Royal Society of London, it developed out of informal gatherings of scientists, including Rene Descartes, Christiaan Huygens, and Pascal.

The Academie was originally intended by Louis XIV (1638-1715) to embrace history and literature as well as science. The Academie was reorganized and given a formal constitution in 1699; in the 18th century it became the leading force in European science.

Since its inception, the Academie has granted its members a state pension, plus financial assistance with their research. The Academie served as a model for the Academy of Sciences of the Russian Federation.

4.2 BRITISH ROYAL SOCIRTY

The Royal Society (of London for Improving Natural Knowledge) is one of the world's oldest and most prestigious scientific societies. The first such society in Britain, it was founded in 1660 as a fellowship of some forty natural philosophers meeting in London, and received its first Royal Charter from Charles II in 1662.

Its *Philosophical Transactions* (1665-) was the first permanent scientific journal. Among its earliest members were Boyle, Hooke, Pepys, Wren, and Newton, whose *Principia* was published in 1687 with the active encouragement of the Society.

Among more literary members were Dryden, Evelyn, and Aubrey. Political and religious topics were excluded from its discussions and debates, and in 1848 the Society became wholly scientific: only those who had made a distinguished contribution to the sciences were eligible for election as Fellows of the Royal Society (FRS).

Today the Royal Society is an independent, self-governing institution for promoting the natural sciences, including mathematics and applied sciences such as engineering and medicine. It operates both nationally and internationally, by encouraging the exchange of scientific ideas through meetings and publications; making grants and research appointments; collaborating with national academies overseas; providing independent advice to the UK government and others; representing the interests of the UK scientific community; and supporting research on the history of science. The Fellowship now numbers around 1,100, including about 100 Foreign Members.

5. ALTERNATIVE TECHNOLOGY

5.1 APPROPRIATE TECHNOLOGY

Appropriate technology (alternative technology, intermediate technology) is an approach to the development and use of technology, dating from the late 1960s, reflecting a concern for minimizing environmental impact and increasing social equity.

A pioneer in this field was the British economist E. F. Schumacher, whose book *Small is Beautiful* (1969) argued against large-scale, centralized technology based on profit considerations, and called for a more human-oriented approach.

Attempts to apply this approach in the Third World have led to the development of an intermediate technology, using local skills and resources, as opposed to capital-intensive high technology, or indigenous low technology.

Intermediate technology has made significant contributions to health care, agricultural practice and food production, manufacturing, and energy production.

Examples include the development of small, locally made water turbines to drive grinding mills and to generate electricity; inexpensive, easily manufactured building materials for low-cost housing; the manufacture of cheap, efficient stoves able to utilize a range of fuels; and improved tools and equipment for agriculture and craft professions.

The organization of collective or co-operative groups has given many poorer people access to otherwise unaffordable technology, for example pumps for irrigation or to ensure water supplies. Somewhat similar concepts applied in Western countries have given rise to the term 'alternative technology', implying an alternative to existing forms of technology.

The initial emphasis of alternative technology was on rural self-sufficiency, often using organic farming practices. Subsequently the Western alternative technology movement, along with the environmental movement, has emphasized renewable energy technology (for example, solar, wind, or tidal energy); and conservation of energy and raw materials through such methods as improved building insulation and recycling.

In the 1980s commercial wind farms were developed in California, USA, and Denmark, and research was undertaken world-wide into

25

solar and geothermal energy, bio-fuel energy crops, and other renewable energy sources. Some developments have moved away from the 'small is beautiful' concept into large-scale high technology. However, given proper funding, renewable energy technology could play an important role in helping to solve many of our perceived energy and environmental problems.

5.2 WIND POWER

Wind power, electric power generated by wind machines. These have the advantage of being relatively safe and a pollution-free method of providing power. It is estimated that wind generators could satisfy up to a fifth of the demand for electric power in many countries, but in 1990 the highest proportion of demand met by wind generation was 1 per cent, in Denmark.

The first wind generator was built in 1890, also in Denmark, starting a 30-year programme there which produced over 100 machines. In the years between the two World Wars cheap fuel everywhere gave little incentive for development of large-scale systems, but thousands of 1 kW machines were built for farms and remote communities.

With a generator mounted behind the propeller, these no longer needed the vertical shaft and gearing of earlier machines. The 1940s saw new interest in generators in the 10-100 kW range, while the Putnam machine in Vermont, USA, 50 m (160 feet) in diameter, produced up to 1,250 kW for a few years.

No larger machine was built until the period following the 1970s oil crises, which brought many two-bladed and three-bladed horizontal-axis turbines, with output from 10 to 3,000 kW. This period also saw the introduction of vertical-axis wind turbines. First proposed in 1931, these have the advantage of direct drive to a ground-level generator and an independence of wind direction.

Their blades form either a bowed 'egg-beater' or a straight 'H' shape and, as in other modern machines, are thin aero-foils. Some people object to the large number of wind machines required to produce significant power in prominent positions along coasts and on hilltops. Alternative locations, such as out at sea, may provide the solution: the winds are stronger and steadier, and larger more productive machines may be constructed without fear of altering existing natural features.

In spite of concern about their location, in 1993 19 UK wind farms (groups of up to several hundred wind generators at one site) have

26

generated sufficient current for 150,000 homes. Similar wind farms exist in the USA.

5.3 RECYCLING

Recycling is the processing of waste so that it can be recovered and reused. Historically, recycling has been done mainly for economic reasons, but recently its value in reducing the depletion of natural resources has been emphasized. Recycling items of domestic refuse also helps reduce the growing problem of waste disposal.

Metals such as scrap steel are returned to the steel mill for reprocessing, though problems are faced when different metals are intimately associated, for example, steel and tin in tin plate. Such 'external recycling' contrasts with 'internal recycling' during an industrial process, for example the re-smelting and recasting of metal turnings and offcuts. Much paper and glass is recycled, the latter exceeding 3 million tonnes every year in Europe alone. Recycling of paper has become important in view of accelerating deforestation.

French, German, Japanese, and Norwegian authorities have taken steps to set up stable markets for recycled paper. Road materials such as asphalt and concrete can be planed off the road, heated, mixed, and re-laid.

Thermoplastics can quite easily be recycled if the waste is a fairly pure sample of one plastic, but recycling of mixed plastics gives poor results. One answer to this problem is the use of compatibilizers, molecules that are added to a mixture of molten plastics and bind them together at a molecular level, to make plastic 'alloys'. Some manufacturers of engineering plastics for the motor car and aerospace industry design plastic products that are easier to recycle.

The technology, legislation, and application of recycling domestic waste is particularly advanced in Japan and Germany. The USA has taken an increasing interest: in some states those who fail to recycle their waste can be prosecuted, and in an increasing number of states domestic waste must be sorted by the householder before collection in order to assist recycling.

Green parties have suggested encouraging recycling by charging a natural resources tax on newly manufactured items: the surcharge on a new car would be returnable when the wreck of the car is delivered for the metal to be recycled. Governments could also charge manufacturers a tax to pay for the disposal of non-renewable packaging.

6. SCIENTIFIC KNOWLEDGE

6.1 APPLICATIONS FOR HUMAN LIFE

Studies for many branches of technology are currently acknowledged as the predominant means for the application of scientific knowledge to the practical aims of human life or, as the description of technology are sometimes phrased, to the change and manipulation of the human environment.

The subject of technology and its branches are treated as the most important subjects in a number of published papers. The contents of the research papers include the description of the materials that are both the object and means of manipulating the environment. Some of these are:

- Industrial Ceramics;
- Industrial Glass;
- Industries, Extraction and Processing;
- Industrial metals;
- Plastics (thermoplastic and thermosetting resins);
- Elastomers (natural and synthetic rubber);
- Energy Conversion;
- Fossil fuels;
- Food production;
- Construction technology;
- Environmental works;
- Transportation;
- Communications technology,
- Computers, Information Processing and Information Systems;
- Photography, Printing, Typography, and Photoengraving;
- Manufacturing;
- Medical applications;
- Military applications;

- Automation;

- Engineering;

6.2 SCIENCE AND TECHNOLOGY

Among the insights that arise from the reviews of the history of technology is the light it throws on the distinction between science and technology.

The history of technology is longer than and distinct from the history of science. Technology is the systematic study of techniques for making and doing things; science is the systematic attempt to understand and interpret the world.

While technology is concerned with the fabrication and use of artefacts, science is devoted to the more conceptual enterprise of understanding the environment, and it depends upon the comparatively sophisticated skills of literacy and numeracy.

Such skills became available only with the emergence of the great world civilizations, so that it is possible to say that science began with those civilizations, some 3,000 years BC, whereas technology, as we have seen, is as old as manlike life.

Science and technology developed as different and separate activities, the former being for several millennia a field of fairly abstruse speculation practiced by a class of aristocratic philosophers, while the latter remained a matter of essentially practical concern to craftsmen of many types.

There were points of intersection, such as the use of mathematical concepts in building and irrigation work, but for the most part the functions of scientist and technologist (to use these modern terms retrospectively) remained distinct in the ancient cultures.

The situation began to change during the medieval period of development in the West (AD 500-1500), when both technical innovation and scientific understanding interacted with the stimuli of commercial expansion and a flourishing urban culture. The robust growth of technology in these centuries could not fail to attract the interest of educated men.

Early in the 17th century, the natural philosopher Francis Bacon had recognized three great technological innovations--the magnetic compass, the printing press, and gunpowder--as the distinguishing achievements of modern man, and he had advocated experimental science as a means of enlarging man's dominion over nature.

29

By emphasizing a practical role for science in this way, Bacon implied a harmonization of science and technology, and he made his intention explicit by urging scientists to study the methods of craftsmen and craftsmen to learn more science. Bacon, with Descartes and other contemporaries, for the first time saw man becoming the master of nature, and a convergence between the traditional pursuits of science and technology was to be the way by which such mastery could be achieved.

Yet the wedding of science and technology proposed by Bacon was not soon consummated. Over the next 200 years, carpenters and mechanics--practical men of long standing--built iron bridges, steam engines, and textile machinery without much reference to scientific principles, while scientists--still amateurs--pursued their investigations in a haphazard manner. But the body of men, inspired by Baconian principles, who formed the Royal Society in London in 1660 represented a determined effort to direct scientific research toward useful ends, first by improving navigation and cartography, and ultimately by stimulating industrial innovation and the search for mineral resources. Similar bodies of scholars developed in other European countries, and by the 19th century scientists were moving toward a professionalism in which many of the goals were clearly the same as those of the technologists.

Thus Justus von Liebig of Germany, one of the fathers of organic chemistry and the first proponent of mineral fertilizer, provided the scientific impulse that led to the development of synthetic dyes, high explosives, artificial fibres, and plastics; and Michael Faraday, the brilliant British experimental scientist in the field of electromagnetism, prepared the ground that was exploited by Thomas A. Edison and many others.

The role of Edison is particularly significant in the deepening relationship between science and technology, because the prodigious trial-and-error process by which he selected the carbon filament for his electric light bulb in 1879 resulted in the creation at Menlo Park, N.J., of what may be regarded as the world's first genuine industrial research laboratory. From this achievement the application of scientific principles to technology grew rapidly.

It led easily to the engineering rationalism applied by Frederick W. Taylor to the organization of workers in mass production, and to the time-and-motion studies of Frank and Lillian Gilbreth at the beginning of the 20th century. It provided a model that was applied rigorously by Henry Ford in his automobile assembly plant and that was followed by

30

every modern mass-production process. It pointed the way to the development of systems engineering, operations research, simulation studies, mathematical modelling, and technological assessment in industrial processes.

This was not just a one-way influence of science on technology, because technology created new tools and machines with which the scientists were able to achieve an ever-increasing insight into the natural world. Taken together, these developments brought technology to its modern highly efficient level of performance

6.3 PUTATIVE AUTONOMY OF TECHNOLOGY

The definition of technology as the systematic study of techniques for making and doing things establishes technology as a social phenomenon and thus as one that cannot possess complete autonomy, unaffected by the society in which it exists. It is necessary to make what may seem to be such an obvious statement because so much autonomy has been ascribed to technology, and the element of despair in interpretations like that of Jacques Ellul is derived from an exaggerated view of the power of technology to determine its own course apart from any form of social control.

Of course it must be admitted that once a technological development, such as the transition from sail to steam power in ships or the introduction of electricity for domestic lighting, is firmly established, it is difficult to stop it before the process is complete. The assembly of resources and the arousal of expectations both create a certain technological momentum that tends to prevent the process from being arrested or deflected.

Nevertheless, the decisions about whether to go ahead with a project or to abandon it are undeniably human, and it is a mistake to represent technology as a monster or a juggernaut threatening human existence. In itself, technology is neutral and passive: in the phrase of Lynn White, Jr., "Technology opens doors; it does not compel man to enter."

Or, in the words of the traditional adage, it is a poor craftsman who blames his tools, and so just as it was naive for the 19th-century optimists to imagine that technology could bring paradise on Earth, it seems equally simplistic for the 20th-century pessimists to make technology itself a scapegoat for man's shortcomings

31

6.4 CRITICISMS OF TECHNOLOGY

Judged entirely on its own traditional grounds of evaluation--that is, in terms of efficiency--the achievement of modern technology has been admirable. Voices from other fields, however, began to raise disturbing questions, grounded in other modes of evaluation, as technology became a dominant influence in society.

In the mid-19th century, the non-technologists were almost unanimously enchanted by the wonders of the new man-made environment growing up around them. London's Great Exhibition of 1851, with its arrays of machinery housed in the truly innovative Crystal Palace, seemed to be the culmination of Francis Bacon's prophetic forecast of man's increasing dominion over nature.

The new technology seemed to fit the prevailing laissez-faire economics precisely and to guarantee the rapid realization of the Utilitarian philosophers' ideal of "the greatest good for the greatest number." Even Marx and Engels espousing a radically different political orientation, welcomed technological progress because in their eyes it produced an imperative need for socialist ownership and control of industry.

Similarly, early exponents of science fiction such as Jules Verne and H.G. Wells explored with zest the future possibilities opened up to the optimistic imagination by modern technology, and the American utopian Edward Bellamy, in his novel *Looking Backward* (1888), envisioned a planned society in the year 2000 in which technology would play a conspicuously beneficial role. Even such late-Victorian literary figures as Lord Tennyson and Rudyard Kipling acknowledged the fascination of technology in some of their images and rhythms.

Yet even in the midst of this Victorian optimism, a few voices of dissent were heard, such as Ralph Waldo Emerson's ominous warning that "Things are in the saddle and ride mankind." For the first time it began to seem as if "things"--the artefacts made by man in his campaign of conquest over nature--might get out of control and come to dominate him.

Samuel Butler, in his satirical novel *Erewhon* (1872), drew the radical conclusion that all machines should be consigned to the scrap heap; and others such as William Morris, with his vision of a reversion to a craft society without modern technology, and Henry James, with his disturbing sensations of being overwhelmed in the presence of modern machinery, began to develop a profound moral critique of the apparent achievements of technologically dominated progress.

32

Even H.G. Wells, despite all the ingenious and prophetic technological gadgetry of his earlier novels, lived to become disillusioned about the progressive character of Western civilization: his last book was entitled *Mind at the End of Its Tether* (1945). Another novelist, Aldous Huxley, expressed disenchantment with technology in a forceful manner in *Brave New World* (1932). Huxley pictured a society of the near future in which technology was firmly enthroned, keeping human beings in bodily comfort without knowledge of want or pain, but also without freedom, beauty, or creativity, and robbed at every turn of a unique personal existence.

An echo of the same view found poignant artistic expression in the film *Modern Times* (1936), in which Charlie Chaplin depicted the depersonalizing effect of the mass-production assembly line. Such images were given special potency by the international political and economic conditions of the 1930s, when the Western world was plunged in the Great Depression and seemed to have forfeited the chance to remould the world order shattered by World War I. In these conditions, technology suffered by association with the tarnished idea of inevitable progress.

Paradoxically, the escape from a decade of economic depression and the successful defence of Western democracy in World War II did not bring a return of confident notions about progress and faith in technology.

The horrific potentialities of nuclear war were revealed in 1945, and the division of the world into hostile power blocs prevented any such euphoria and served to stimulate criticisms of technological aspirations even more searching than those that have already been mentioned. J. Robert Oppenheimer, who directed the design and assembly of the atomic bombs at Los Alamos, N.M., later opposed the decision to build the thermonuclear (fusion) bomb and described the accelerating pace of technological change with foreboding: "One thing that is new is the prevalence of newness, the changing scale and scope of change itself, so that the world alters as we walk in it, so that the years of man's life measure not some small growth or rearrangement or moderation of what he learned in childhood, but a great upheaval."

The theme of technological tyranny over man's individuality and his traditional patterns of life were expressed by Jacques Ellul, of the University of Bordeaux, in his book *The Technological Society* (1964, first published as *La Technique* in 1954). Ellul asserted that technology had become so pervasive that man now lived in a milieu of technology rather than of nature. He characterized this new milieu as artificial, autonomous, self-determining, nihilistic (that is, not directed to ends, though proceeding by cause and effect), and, in fact, with means enjoying primacy over ends.

Technology, Ellul held, had become so powerful and ubiquitous that other social phenomena such as politics and economics had become situated *in it* rather than influenced *by it*. The individual, in short, had come to be adapted to the technical milieu rather than the other way round.

While views such as those of Ellul have enjoyed a considerable vogue since World War II, and have spawned a remarkable subculture of "hippies" and others who have sought, in a variety of ways, to reject participation in technological society, it is appropriate to make two observations on them. The first is that these views are, in a sense, a luxury enjoyed only by advanced societies, which have benefited from modern technology.

Few voices critical of technology can be heard in developing countries that are hungry for the advantages of greater productivity and the rising standards of living that have been seen to accrue to technological progress in the more fortunate developed countries.

Indeed, the anti-technological movement is greeted with complete incomprehension in these parts of the world, so that it is difficult to avoid the conclusion that only when the whole world enjoys the benefits of technology can we expect the more subtle dangers of technology to be appreciated, and by then, of course, it may be too late to do anything about them.

The second observation about the spate of technological pessimism in the advanced countries is that it has not managed to slow the pace of technological advance, which seems, if anything, to have accelerated in the 20th century. The gap between the first powered flight and the first human steps on the Moon was only 66 years, and that between the disclosure of the fission of uranium and the detonation of the first atomic bomb was a mere six and a half years.

The advance of the information revolution based on the electronic computer has been exceedingly swift, so that despite the denials of the possibility by elderly and distinguished experts, the sombre spectre of sophisticated computers replicating higher human mental functions and even human individuality should not be relegated too hurriedly to the classification of science fantasy.

The bio-technic stage of technological innovation is still in its infancy, and if the recent rate of development is extrapolated forward many seemingly impossible targets could be achieved in the next century. Not that this will be any consolation to the pessimists, as it only indicates the ineffectiveness to date of attempts to slow down technological progress

7. ADVANCED TECHNOLOGY

7.1 NUCLEAR TECHNOLOGY

The solution to the first problem that of controlling nuclear technology, is primarily political. At its root is the anarchy of national self-government, for as long as the world remains divided into a multiplicity of nation-states, or even into two power blocs, each committed to the defence of its own sovereign power to do what it chooses, nuclear weapons merely replace the older weapons by which such nation-states have maintained their independence in the past.

The availability of a nuclear armoury has emphasized the weaknesses of a world political system based upon sovereign nation-states. Here, as elsewhere, technology is a tool that can be used creatively or destructively. But the manner of its use depends entirely on human decisions, and in this matter of nuclear self-control the decisions are those of governments.

There are other aspects of the problem of nuclear technology, such as the disposal of radioactive waste and the quest to harness the energy released by fusion, but although these are important issues in their own right, they are subordinate to the problem of the use of nuclear weapons in warfare

7.2 MILITARY TECHNOLOGY

One area of technology was not dramatically influenced by the application of steam or electricity by the end of the 19th century: military technology. Although the size of armies increased between 1750 and 1900, there were few major innovations in techniques, except at sea where naval architecture rather reluctantly accepted the advent of the iron steamship and devoted itself to matching ever-increasing firepower with the strength of the armour plating on the hulls.

The quality of artillery and of firearms improved with the new high explosives that became available in the middle of the 19th century, but experiments such as the three-wheeled iron gun carriage, invented by the French army engineer Nicolas Cugnot in 1769, which counts as the first steam-powered road vehicle, did not give rise to any confidence that steam could be profitably used in battle.

Railroads and the electric telegraph were put to effective military use, but in general it is fair to say that the 19th century put remarkably little of its tremendous and innovative technological effort into devices for war.

In the course of its dynamic development between 1750 and 1900, important things happened to technology itself. In the first place, it became self-conscious. This change is sometimes characterized as one from a craft-based technology to one based on science, but this is an oversimplification.

What occurred was rather an increase in the awareness of technology as a socially important function. It is apparent in the growing volume of treatises on technological subjects from the 16th century onward and in the rapid development of patent legislation to protect the interests of technological innovators.

It is apparent also in the development of technical education, uneven at first, being confined to the French polytechnics and spreading thence to Germany and North America but reaching even Britain, which had been most opposed to its formal recognition as part of the structure of education, by the end of the 19th century. Again, it is apparent in the growth of professional associations for engineers and for other specialized groups of technologists.

Second, by becoming self-conscious, technology attracted attention in a way it had never done before, and vociferous factions grew up to praise it as the mainspring of social progress and the development of democracy or to criticize it as the bane of modern man, responsible for the harsh discipline of the "dark Satanic mills" and the tyranny of the machine and the squalor of urban life.

It was clear by the end of the 19th century that technology was an important feature in industrial society and that it was likely to become more so. Whatever was to happen in the future, technology had come of age and had to be taken seriously as a formative factor of the utmost significance in the continuing development of civilization

36

8. MODERN TECHNOLOGY

8.1 TECHNOLOGICAL DILEMMA

Whatever the responses to modern technology, there can be no doubt that it presents contemporary society with a number of immediate problems that take the form of a traditional choice of evils, so that it is appropriate to regard them as constituting a "technological dilemma."

This is the dilemma between, on the one hand, the overdependence of life in the advanced industrial countries on technology, and, on the other hand, the threat that technology will destroy the quality of life in modern society and even endanger society itself.

Technology thus confronts Western civilization with the need to make a decision, or rather, a series of decisions, about how to use the enormous power available to society constructively rather than destructively. The need to control the development of technology, and so to resolve the dilemma, by regulating its application to creative social objectives, makes it ever more necessary to define these objectives while the problems presented by rapid technological growth can still be solved.

These problems, and the social objectives related to them, may be considered fewer than three broad headings. First is the problem of controlling the application of nuclear technology. Second is the population problem, which is twofold: it seems necessary to find ways of controlling the dramatic rise in the number of human beings and, at the same time, to provide food and care for the people already living on the Earth. Third, there is the ecological problem, whereby the products and wastes of technical processes have polluted the environment and disturbed the balance of natural forces of regeneration.

When these basic problems have been reviewed it will be possible, finally, to consider the effect of technology on life in town and countryside, and to determine the sort of judgments about technology and society to which a study of the history of technology leads.

8.2 POWER TECHNOLOGY

An outstanding feature of the Industrial Revolution has been the advance in power technology. At the beginning of this period, the major sources of power available to industry and any other potential consumer were animate energy and the power of wind and water, the

37

only exception of any significance being the atmospheric steam engines that had been installed for pumping purposes, mainly in coal mines.

It is to be emphasized that this use of steam power was exceptional and remained so for most industrial purposes until well into the 19th century. Steam did not simply replace other sources of power: it transformed them. The same sort of scientific inquiry that led to the development of the steam engine was also applied to the traditional sources of inanimate energy, with the result that both waterwheels and windmills were improved in design and efficiency.

Numerous engineers contributed to the refinement of waterwheel construction, and by the middle of the 19th century new designs made possible increases in the speed of revolution of the waterwheel and thus prepared the way for the emergence of the water turbine, which is still an extremely efficient device for converting energy.

8.3 SPACE AGE TECHNOLOGY

The years since World War II ended have been spent in the shadow of nuclear weapons, even though they have not been used in war since that time. These weapons have undergone momentous development: the fission bombs of 1945 were superseded by the more powerful fusion bombs in 1950, and before 1960 rockets were shown capable of delivering these weapons at ranges of thousands of miles.

This new military technology has had an incalculable effect on international relations, for it has contributed to the polarization of world power blocs while enforcing a caution, if not discipline, in the conduct of international affairs that was absent earlier in the 20th century.

The fact of nuclear power has been by no means the only technological novelty of the post-1945 years. So striking, indeed, have been the advances in engineering, chemical and medical technology, transport, and communications, which some commentators have written, somewhat misleadingly, of the "second Industrial Revolution" in describing the changes in these years.

The rapid development of electronic engineering has created a new world of computer technology, remote control, miniaturization, and instant communication. Even more expressive of the character of the period has been the leap over the threshold of extraterrestrial exploration. The techniques of rocketry, first applied in weapons, were developed to provide launch vehicles for satellites and lunar and planetary probes and eventually, in 1969, to set the first men on the Moon and to bring them home safely again.

This astonishing achievement was stimulated in part by the international ideological rivalry already mentioned, as only the Soviet Union and the United States had both the resources and the will to support the huge expenditures required. It justifies the description of this period, however, as that of "space age technology."

8.4 EDUCATION AND TECHNOLOGY

A third theme to emerge from the review of the history of technology is the growing importance of education. In the early millennia of human existence, a craft was acquired in a lengthy and laborious manner by serving with a master who gradually trained the initiate in the arcane mysteries of the skill.

Such instruction, set in a matrix of oral tradition and practical experience, was frequently more closely related to religious ritual than to the application of rational scientific principles. Thus the artisan in ceramics or sword making protected the skill while ensuring that it would be perpetuated.

Craft training was institutionalized in Western civilization in the form of apprenticeship, which has survived into the 20th century as a framework for instruction in technical skills. Increasingly, however, instruction in new techniques has required access both to general theoretical knowledge and to realms of practical experience that, on account of their novelty, were not available through traditional apprenticeship.

Thus the requirement for a significant proportion of academic instruction has become an important feature of most aspects of modern technology. This has accelerated the convergence between science and technology in the 19th and 20th centuries and has created a complex system of educational awards representing the level of accomplishment from simple instruction in schools to advanced research in universities.

French and German academies led in the provision of such theoretical instruction, while Britain lagged somewhat in the 19th century, owing to its long and highly successful tradition of apprenticeship in engineering and related skills. But by the 20th century all the advanced industrial countries, including newcomers like Japan, had recognized the crucial role of a theoretical technological education in achieving commercial and industrial competence.

39

The recognition of the importance of technological education, however, has never been complete in Western civilization, and the continued coexistence of other traditions has caused problems of assimilation and adjustment.

The British author C.P. Snow drew attention to one of the most persistent problems in his perceptive essay *The Two Cultures* (1959), in which he identified the dichotomy between scientists and technologists on the one hand and humanists and artists on the other as one between those who did understand the second law of thermodynamics and those who did not, causing a sharp disjunction of comprehension and sympathy.

Arthur Koestler put the same point in another way by observing that the traditionally humanities-educated Western man is reluctant to admit that a work of art is beyond his comprehension, but will cheerfully confess that he does not understand how his radio or heating system works. Koestler characterized such a modern man, isolated from a technological environment that he possesses without understanding, as an "urban barbarian."

Yet the growing prevalence of "black-box" technology, in which only the rarefied expert is able to understand the enormously complex operations that go on inside the electronic equipment, makes it more and more difficult to avoid becoming such a "barbarian." The most helpful development would seem to be not so much seeking to master the expertise of others in our increasingly specialized society, as encouraging those disciplines that provide bridges between the two cultures, and here there is a valuable role for the history of technology.

8.5 EMERGENCE OF WESTERN TECHNOLOGY

The technological history of the Middle Ages was one of slow but substantial development. In the succeeding period the tempo of change increased markedly and was associated with profound social, political, religious, and intellectual upheavals in Western Europe.

The emergence of the nation-state, the cleavage of the Christian Church by the Protestant Reformation, the Renaissance and its accompanying scientific revolution, and the overseas expansion of European states all had interactions with developing technology. This expansion became possible after the advance in naval technology opened up the ocean routes to Western navigators.

The conversion of voyages of discovery into imperialism and colonization was made possible by the new firepower. The combination

Andreas Sofroniou

of light, manoeuvrable ships with the firepower of iron cannon gave European adventurers a decisive advantage, enhanced by other technological assets.

The Reformation, not itself a factor of major significance to the history of technology, nevertheless had interactions with it: the capacity of the new printing presses to disseminate all points of view contributed to the religious upheavals, while the intellectual ferment provoked by the Reformation resulted in a rigorous assertion of the vocational character of work and thus stimulated industrial and commercial activity and technological innovation.

It is an indication of the nature of this encouragement that so many of the inventors and scientists of the period were Calvinists, Puritans, and, in England, Dissenters

8.6 INTERNATIONAL TRADE TECHNOLOGY

Technological development can also provide a distinctive trade advantage. The relatively advanced countries--particularly the United States, Japan, and those of Western Europe--are the principal exporters of high-technology products such as computers and precision machinery.

One important aspect of technology is that it can change rapidly. This is perhaps most obvious in the computer field, where productivity has increased, and costs have fallen sharply since the early 1960s. Such rapid changes present several challenges. For the countries that are not in the front rank, it raises the question of whether they should import high-technology products or attempt to enter the circle of the most advanced nations.

For the countries that have held the technological lead in the past, there is always the possibility that they will be successfully overtaken by newcomers. This occurred in the second half of the 20th century when Japan advanced technologically in its automobile production to the point where it could challenge the automobile leadership of North America and Europe. Japan quickly became the world's foremost producer of automobiles.

8.7 TECHNOLOGICAL SOCIETY

Much of the 19th-century optimism about the progress of technology has dispersed, and an increasing awareness of the technological dilemma confronting the world makes it possible to offer a realistic

41

assessment of the role of technology in shaping society at the end of the 20th century.

In the first place, it can be clearly recognized that the relationship between technology and society is complex. Any technological stimulus can trigger a variety of social responses, depending on such unpredictable variables as differences between human personalities; similarly, no specific social situation can be relied upon to produce a determinable technological response.

Any "theory of invention," therefore, must remain extremely tentative, and any notion of a "philosophy" of the history of technology must allow for a wide range of possible interpretations. A major lesson of the history of technology, indeed, is that it has no precise predictive value. It is frequently possible to see in retrospect when one particular artefact or process had reached obsolescence while another promised to be a highly successful innovation, but at the time such historical hindsight is not available and the course of events is indeterminable.

In short, the complexity of human society is never capable of resolution into a simple identification of causes and effects driving historical development in one direction rather than another and any attempt to identify technology as an agent of such a process is unacceptable.

Much of the 19th-century optimism about the progress of technology has dispersed, and an increasing awareness of the technological dilemma confronting the world makes it possible to offer a realistic assessment of the role of technology in shaping society at the end of the 20th century.

8.8 SOCIAL INVOLVEMENT IN TECHNOLOGY

An awareness of this interaction is important in surveying the development of technology through successive civilizations. To simplify the relationship as much as possible, there are three points at which there must be some social involvement in technological innovation: social need, social resources, and a sympathetic social ethos. In default of any of these factors it is unlikely that a technological innovation will be widely adopted or be successful.

The sense of social need must be strongly felt, or people will not be prepared to devote resources to a technological innovation. The thing needed may be a more efficient cutting tool, a more powerful lifting device, a laboursaving machine, or a means of utilizing new fuels or a new source of energy. Or, because military needs have always provided a stimulus to technological innovation, it may take the form of a

42

requirement for better weapons. In modern societies, needs have been generated by advertising. Whatever the source of social need, it is essential that enough people be conscious of it to provide a market for an artefact or commodity that can meet the need.

Social resources are similarly an indispensable prerequisite to a successful innovation. Many inventions have foundered because the social resources vital for their realization--the capital, materials, and skilled personnel--were not available. The notebooks of Leonardo da Vinci are full of ideas for helicopters, submarines, and airplanes, but few of these reached even the model stage because resources of one sort or another were lacking.

The resource of capital involves the existence of surplus productivity and an organization capable of directing the available wealth into channels in which the inventor can use it. The resource of materials involves the availability of appropriate metallurgical, ceramic, plastic, or textile substances that can perform whatever functions a new invention requires of them. The resource of skilled personnel implies the presence of technicians capable of constructing new artifacts and devising novel processes. A society, in short, has to be well primed with suitable resources in order to sustain technological innovation.

A sympathetic social ethos implies an environment receptive to new ideas, one in which the dominant social groups are prepared to consider innovation seriously. Such receptivity may be limited to specific fields of innovation--for example, improvements in weapons or in navigational techniques--or it may take the form of a more generalized attitude of inquiry, as was the case among the industrial middle classes in Britain during the 18th century, who were willing to cultivate new ideas and inventors, the breeders of such ideas. Whatever the psychological basis of inventive genius, there can be no doubt that the existence of socially important groups willing to encourage inventors and to use their ideas has been a crucial factor in the history of technology.

Social conditions are thus of the utmost importance in the development of new techniques, some of which will be considered below in more detail. It is worthwhile, however, to register another explanatory note. This concerns the rationality of technology. It has already been observed that technology involves the application of reason to techniques, and in the 20th century it has come to be regarded as almost axiomatic that technology is a rational activity stemming from the traditions of modern science.

43

Nevertheless, it should be observed that technology, in the sense in which the term is being used here, is much older than science, and also that techniques have tended to ossify over centuries of practice or to become diverted into such para-rational exercises as alchemy.

Some techniques became so complex, often depending upon processes of chemical change that were not understood even when they were widely practiced, that technology sometimes became itself a "mystery" or cult into which an apprentice had to be initiated like a priest into holy orders, and in which it was more important to copy an ancient formula than to innovate.

The modern philosophy of progress cannot be read back into the history of technology; for most of its long existence technology has been virtually stagnant, mysterious, and even irrational. It is not fanciful to see some lingering fragments of this powerful technological tradition in the modern world, and there is more than an element of irrationality in the contemporary dilemma of a highly technological society contemplating the likelihood that it will use its sophisticated techniques in order to accomplish its own destruction.

It is thus necessary to beware of over-facile identification of technology with the "progressive" forces in contemporary civilization.

On the other hand it is impossible to deny that there is a progressive element in technology, as it is clear from the most elementary survey that the acquisition of techniques is a cumulative matter, in which each generation inherits a stock of techniques on which it can build if it chooses and if social conditions permit.

Over a long period of time the history of technology inevitably highlights the moments of innovation that show this cumulative quality as some societies advance, stage by stage, from comparatively primitive to more sophisticated techniques.

But although this development has occurred and is still going on, it is not intrinsic to the nature of technology that such a process of accumulation should occur, and it has certainly not been an inevitable development.

The fact that many societies have remained stagnant for long periods of time, even at quite developed stages of technological evolution, and that some have actually regressed and lost the accumulated techniques passed on to them, demonstrates the ambiguous nature of technology and the critical importance of its relationship with other social factors.

9 FIELDS OF TECHNOLOGY

9.1 VARIOUS OTHER ASPECTS

In manufacturing, transport, and military technology, the achievements of the Greco-Roman period are not remarkable. The major manufacturing crafts--the making of pottery and glass, weaving, leatherworking, fine-metalworking, and so on--followed the lines of previous societies, albeit with important developments in style.

Superbly decorated Athenian pottery, for example, was widely dispersed along the trade routes of the Mediterranean, and the Romans made good quality pottery available throughout their empire through the manufacture and trade of the standardized red ware called terra sigillata, which was produced in large quantities at several sites in Italy and Gaul.

9.2 TRANSPORT

Transport, again, followed earlier precedents, the sailing ship emerging as a seagoing vessel with a carvel-built hull (that is, with planks meeting edge-to-edge rather than overlapping as in clinker-built designs), and a fully developed keel with stem-post and sternpost.

The Greek sailing ship was equipped with a square or rectangular sail to receive a following wind and one or more banks of oarsmen to propel the ship when the wind was contrary. The Greeks began to develop a specialized fighting ship, provided with a ram in the prow, and the cargo ship, dispensing with oarsmen and relying entirely upon the wind, was also well established by the early years of classical Greece.

The Romans took over both forms, but without significant innovation. They gave much more attention to inland transport than to the sea, and they constructed a remarkable network of carefully aligned and well-laid roads, often paved over long stretches, throughout the provinces of the empire.

Along these strategic highways the legions marched rapidly to the site of any crisis at which their presence was required. The roads also served for the development of trade, but their primary function was always military, as a vital means of keeping a vast empire in subjection.

45

9.3 MOTION-PICTURE TECHNOLOGY

Technology is the major means for the production and showing of motion pictures. It includes not only the motion-picture camera and projector but also such technologies as those involved in recording sound, in editing both picture and sound, in creating special effects, and in producing animation.

Motion-picture technology is a curious blend of the old and the new. In one piece of equipment state-of-the-art digital electronics may be working in tandem with a mechanical system invented in 1895.

Furthermore, the technology of motion pictures is based not only on the prior invention of still photography but also on a combination of several more or less independent technologies; that is, camera and projector design, film manufacture and processing, sound recording and reproduction, and lighting and light measurement.

9.4 PHOTOGRAPHY, AUTOMATIC-DIAPHRAGM SYSTEMS

On a camera with a viewing screen (view camera or single-lens reflex) viewing and focusing are carried out with the lens diaphragm fully open, but the exposure is often made at a smaller aperture. Reflex cameras (and increasingly also view cameras) therefore incorporate a mechanism that automatically or semi-automatically stops down (reduces) the lens to the working aperture immediately before the exposure.

9.5 POLICE TECHNOLOGY VOICE IDENTIFICATION

The use of the sound spectrograph for voice identification is a relatively new development. The sound spectrograph is an instrument that graphically presents the time, frequency, and intensity of speech sound waves. The graphic forms for a sound, as spoken by one person, can be compared with those produced by the speech of a second person and thus differentiated. The accuracy of the technique in identifying individuals remains in doubt among speech scientists, even though voice-graphs, or voice-prints, have been used in court.

9.6 DISTRIBUTION THEORY

Another dynamic influence is technological progress. The concept of the production function assumes a constant technology. But in reality the growth of production is much less the consequence of increased quantities of labour and capital than of improvements in their quality.

This element in increased production is distributed in a way not fully explained by neoclassical theory.

Part of the change in distribution that is caused by technological progress can be analyzed as resulting from changes in the elasticities of production.

Technological change is said to be "capital-using," and the share of capital will increase. This is what, in fact, may have happened; the change in technology has offset, though it has not neutralized, the decline in the share of capital caused by the employment of a higher amount of capital per worker. But another part of the fruits of technological progress is garnered by profit receivers, probably quite a substantial part.

Businessmen who are quick innovators make high profits; in a rapidly changing society, profits tend to be high, a circumstance that is fortunate because profits are the mainspring of economic change. The high rate of growth experienced by the post-World War I Western world stemmed from this profit-innovation-profit nexus.

9.7 ELECTRON BEAM

This is the stream of electrons (as from a betatron) used chiefly in research, technology, and medical therapy to produce X rays and images on television, oscilloscope, and electron-microscope screens. Electrons may be collimated by holes and slits, and, because they are electrically charged, they may be deflected, focused, and energized by electric and magnetic fields

9.8 STRINGED INSTRUMENT AND ELECTRONICS

In the second half of the 20th century, electronic technology has made remarkable changes in the structure and function of many stringed instruments by making amplification and tonal change possible. The best known and most pervasive example is the electric guitar, which, strictly speaking, may be considered a chordophone but is often classified as an electrophone.

The electric guitar may be hollow-bodied like a traditional guitar or solid-bodied, but in either case amplification of the strings is provided by a "pickup" (or contact microphone) that creates artificial resonance through its connection to amplifiers and loudspeakers.

Pickups are often attached to violins, lutes, and other instruments, as well as to guitars, making these instruments usable in noisy

47

environments and vast amphitheatres. Musicians who use such instruments (especially electric guitars) have developed feedback and other techniques that can alter the timbre beyond recognition.

9.9 INFLUENCE OF TECHNOLOGY ON ART

To varying degrees, the arts depend on technological evolution for the very techniques used to create works of art--literature least, music considerably, and architecture most of all. The cinema has been made possible only by recent technological developments.

The arts also depend on technology for the dissemination of the creative product. Book printing has made possible the development of a mass reading public, which in turn has facilitated the rise of new literary genres, such as the novel.

Modern techniques rendered objects of visual art mechanically reproducible, hence perceived and treated as less "unique" than they had been in the past. The film is a peculiar art in that the technique of production and the technique of dissemination immediately imply each other, both having been produced by the same technological development.

Beyond these direct technological influences on the arts, there is an indirect one, mediated through the effects that particular technologies have on the imagination of artists or of larger populations.

9.10 PHOTOGRAPHY, TECHNOLOGY OF DISK FILM

Some compact mass-market cameras take circular disks of film, 65 millimetres in diameter, in light-tight cartridges and coated on a 0.18-mm polyester base. In the camera the disk rotates as up to 15 exposures (frame size 8×10 millimetres) are recorded around the disk circumference. The disk lies flatter in the camera than rolled-up film and is suitable for more automated photofinishing; the high printing magnification required, however, limits the image quality.

9.11 POLICE TECHNOLOGY TAGGING AND TRACKING

Object tagging and signal tracking are a third category of police technical-surveillance operations. Devices emitting a unique signal are attached to vehicles or implanted in targeted objects. Compatible receiving or activating equipment is used to track the tagged objects.

A typical device used in tracking is a miniature radio transmitter using distinctive tone modulation, commonly known as a "bumper beeper," which is surreptitiously attached to a vehicle. The target is discreetly tracked with one or more surveillance vehicles equipped with appropriate receiving equipment.

9.12 ROLE OF POPULATION AND INDUSTRY

Uses of lands and resources are being modified in the expectation of continued population growth, industrial expansion, and accelerating technological change. Yet it is possible that, in the future, uses of lands and resources will take place in times of population stability, little industrial expansion, and a technology directed toward reorganization and a rearrangement of activities to achieve a better environmental relationship. Even though certain countries of the world have already reached some degree of population stability--*e.g.*, Ireland, Hungary, France, Sweden, Switzerland, and Japan--industrial expansion and rapid technological change continue in these countries, in part because of the demands made by other expanding nations.

The existing expansionist phase of technological civilization cannot, however, be expected to continue indefinitely. The ecological limitations on growth in a limited space with limited resources lead to predictions of an inevitable end to this expansion, even if mankind fails to voluntarily limit its own growth.

9.13 VIDEO-TELEPHONE TECHNOLOGY

Future videophone and videoconferencing technology is likely to develop in two different directions. The first direction is toward lower data rates. For example, the International Telecommunication Union has set a goal of establishing standards for low-bit-rate videophone communication in the target range of 9.6 to 28.8 kilobits per second.

The other direction is the development of very-high-quality videoconferencing, so that a goal of conducting business meetings without requiring travel may be reached. In 1990 Bellcore developed a teleconferencing system in which the video images of remote participants appear to be sitting on the opposite side of the room.

This system, known as Video-Window, employs multiple cameras, multiple rear-projection video displays, and high-quality, full-duplex, hands-free communications. Video-Window requires a significant amount of transmission bandwidth in order to support the communication signal provided by the system.

49

9.14 LAMINATION

This is the process of building up successive layers of a substance, such as wood or textiles, and bonding them with resin to form a finished product. Laminated board, for example, consists of thin layers of wood bonded together; similarly, laminated fabric consists of two or more layers of cloth joined together with an adhesive, or a layer of fabric bonded to a plastic sheet.

50

10. HISTORICAL EVENTS

10.1 QUALITY OF LIFE

This theme, concerned with the quality of life, which can be identified in the relationship between technology and society. There can be little doubt that technology has brought a higher standard of living to people in advanced countries, just as it has enabled a rapidly rising population to subsist in the developing countries.

It is the prospect of rising living standards that makes the acquisition of technical competence so attractive to these countries. But however desirable the possession of a comfortable sufficiency of material goods, and the possibility of leisure for recreative purposes, the quality of a full life in any human society has other even more important prerequisites, such as the possession of freedom in a law-abiding community and of equality before the law.

These are the traditional qualities of democratic societies, and it has to be asked whether technology is an asset or a liability in acquiring them. Certainly, highly illiberal regimes have used technological devices to suppress individual freedom and to secure obedience to the state: the nightmare vision of George Orwell's *Nineteen Eighty-four* (1949), with its tele-screens and sophisticated torture, has provided literary demonstration of this reality, should one be needed.

But the fact that high technological competence requires, as has been shown, a high level of educational achievement by a significant proportion of the community holds out the hope that a society that is well-educated will not long endure constraints on individual freedom and initiative that are not self-justifying.

In other words, the high degree of correlation between technological success and educational accomplishment suggests a fundamental democratic bias about modern technology. It may take time to become effective, but given sufficient time without a major political or social disruption and a consequent resurgence of national assertiveness and human selfishness, there are sound reasons for hoping that technology will bring the people of the world into a closer and more creative community.

Such, at least, must be the hope of anybody who takes a long view of the history of technology as one of the most formative and persistently creative themes in the development of mankind from the Palaeolithic cave dwellers of antiquity to the dawn of the space age in the 21st century.

Above all other perceptions of technology, the threshold of space exploration on which mankind stands at the beginning of the 21st century provides the most dynamic and hopeful portent of human potentialities.

Even while the threat of technological self-destruction remains ominous, and the problems of population control and ecological imbalance cry out for satisfactory solutions, man has found a clue of his own future in terms of a quest to explore and colonize the depths of an infinitely fascinating universe.

As yet, only a few visionaries have appreciated the richness of this possibility, and their projections are too easily dismissed as nothing more than imaginative science fiction. But in the long run, if there is to be a long run for our uniquely technological but wilful species, the future depends upon the ability to acquire such a cosmic perspective, so it is important to recognize this now and to begin the arduous mental and physical preparations accordingly.

The words of Arthur C. Clarke, one of the most perceptive of contemporary seers, in his *Profiles of the Future* (1962), are worth recalling in this context. Thinking ahead to the countless aeons that could stem from the remarkable human achievement summarized in the history of technology, he surmised that the all-knowing beings who may evolve from these humble beginnings may still regard our own era with wistfulness: "But for all that, they may envy us, basking in the bright afterglow of Creation; for we knew the Universe when it was young."

10.2 ATOMIC POWER, HISTORY OF

Until 1945, electricity and the internal-combustion engine were the dominant sources of power for industry and transport in the 20th century, although in some parts of the industrialized world steam power and even older prime movers remained important.

Early research in nuclear physics was more scientific than technological, stirring little general interest. In fact, from the work of Ernest Rutherford, Albert Einstein, and others to the first successful experiments in splitting heavy atoms in Germany in 1938, no particular thought was given to engineering potential.

The war led to the Manhattan Project to produce the fission bomb that was first exploded at Alamogordo. Only in its final stages did even this program become a matter of technology, when the problems of building large reactors and handling radioactive materials had to be solved; and at this point it also became an economic and political matter, because very heavy capital expenditure was involved.

Thus, in this crucial event of the mid-20th century, the convergence of science, technology, economics, and politics finally took place

10.3 CIVIL ENGINEERING, HISTORY OF

One industry that has not been deeply influenced by new control-engineering techniques is construction, in which the nature of the tasks involved makes dependence on a large labour force still essential, whether it be in constructing a skyscraper, a new highway, or a tunnel. Nevertheless, some important new techniques have appeared since 1945, notably the use of heavy earth-moving and excavating machines such as the bulldozer and the tower crane.

The use of prefabricated parts according to a predetermined system of construction has become widespread. In the construction of housing units, often in large blocks of apartments or flats, such systems are particularly relevant because they make for standardization and economy in plumbing, heating, and kitchen equipment. The revolution in home equipment that had begun before World War II has continued apace since, with a proliferation of electrical equipment.

10.4 VACUUM TECHNOLOGY SORPTION PUMP

Typically, the size of these pumps is about 1,000 grams of sorbent material, which retains gas molecules on its surface. They are capable of pumping from atmosphere to 10^{-2} torr or can be used in series down to 10^{-5} torr. In most cases the sorbent material is a molecular sieve-- that is, a material that has been processed so that it is porous, with pore sizes comparable to the size of molecules, although activated charcoal can also be employed.

The sorbent is positioned inside a cylindrical container that is connected to the vacuum system and that can be immersed in liquid nitrogen for super-cooling to aid the sorption process. The gas is released when the sorbent returns to room temperature. This pump is used mainly for roughing systems in which the sputter ion and titanium sublimation pumps serve to ensure freedom from organic contamination.

10.5 INTEGRATED CIRCUIT MANUFACTURING

53

The fabrication of ICs begins with the preparation of silicon of very high purity. Single-crystal boules with a diameter as large as 20 centimetres are produced. The boules are sliced into wafers of a specified crystal orientation. When their surfaces have been polished flat to a mirror-like finish that is free of defects, the silicon wafers are ready for fabrication into IC devices.

This manufacturing process typically involves a sequence of more than 600 steps. These steps may be categorized according to function: deposition of thin films, introduction of impurities (doping), lithographic patterning of IC features corresponding to those of the physical layout, etching to define the features of individual circuit elements, annealing to cause a chemical reaction or to form the desired microstructure of deposited films or interfaces between films, polishing to planate the surface, and cleaning to prevent contamination and the consequent introduction of defects into circuit elements by particulate matter. These operations are repeated again and again in different sequential order until the IC fabrication is completed.

10.6 CONSERVATION - TECHNOLOGICAL ISSUES

Technological progress is another important reason for many conflicts in matters pertaining to conservation. Although technology can be a boon, it can also be poorly related to environmental realities. Through the use of technology great environmental changes can quickly be brought about.

Although these changes are usually intended to be beneficial, they frequently occur in natural environments in which all things are ecologically related to one another. As a consequence, the changes may produce side effects that were not anticipated or that were discounted as being of little importance, thereby disrupting other human activities or the environment as a whole.

Examples of such situations include the polluting effects of certain industries or the spread of waterborne diseases following the construction of major irrigation projects.

10.7 URBAN MANUFACTURING, HISTORY OF

Manufacturing industry in the early civilizations concentrated on such products as pottery, wines, oils, and cosmetics, which had begun to circulate along the incipient trade routes before the introduction of metals; these became the commodities traded for the metals. In pottery, the potter's wheel became widely used for spinning the clay into the desired shape, but the older technique of building pots by hand from rolls of clay remained in use for some purposes.

In the production of wines and oils various forms of press were developed, while the development of cooking, brewing, and preservatives justified the assertion that the science of chemistry began in the kitchen. Cosmetics too were an offshoot of culinary art.

Pack animals were still the primary means of land transport, the wheeled vehicle developing slowly to meet the divergent needs of agriculture, trade, and war. In the latter category, the chariot appeared as a weapon, even though its use was limited by the continuing difficulty of harnessing a horse. Military technology brought the development of metal plates for armour.

10.8 BYZANTIUM, HISTORY OF

The immediate eastern neighbour of the new civilization of medieval Europe was Byzantium, the surviving bastion of the Roman Empire based in Constantinople, which endured for 1,000 years after the collapse of the western half of the empire. There the literature and traditions of Hellenic civilization were perpetuated, becoming increasingly available to the curiosity and greed of the West through the traders who arrived from Venice and elsewhere.

Apart from the influence on Western architectural style of such Byzantine masterpieces as the great domed structure of Hagia Sophia, the technological contribution of Byzantium itself was probably slight, but it served to mediate between the West and other civilizations one or more stages removed, such as the Islamic world, India, and China.

10.9 GAS-TURBINE ENGINE, HISTORY OF

The principle of the gas turbine is that of compressing and burning air and fuel in a combustion chamber and using the exhaust jet from this process to provide the reaction that propels the engine forward. In its turbo-propeller form, which developed only after World War II, the exhaust drives a shaft carrying a normal airscrew (propeller). Compression is achieved in a gas-turbine engine by admitting air through a turbine rotor.

In the so-called ramjet engine, intended to operate at high speeds, the momentum of the engine through the air achieves adequate compression. The gas turbine has been the subject of experiments in road, rail, and marine transport, but for all purposes except that of air transport its advantages have not so far been such as to make it a viable rival to traditional reciprocating engines.

10.10 FOOD PRODUCTION, HISTORY OF

Food production has also been subject to technological innovation, such as accelerated freeze-drying and irradiation as methods of preservation, as well as the increasing mechanization of farming throughout the world. The widespread use of new pesticides and herbicides has in some cases reached the point of abuse, causing worldwide concern.

Despite such problems, farming has been transformed in response to the demand for more food; scientific farming, with its careful breeding, controlled feeding, and mechanized handling, has become commonplace.

New food-producing techniques such as aquaculture and hydroponics, for farming the sea and the seabed and for creating self-contained cycles of food production without soil, respectively, are being explored, either to increase the world supply of food or to devise ways of sustaining closed communities such as may one day venture forth from the Earth on the adventure of interplanetary exploration.

10.11 BUILDING TECHNIQUES, HISTORY OF

Prehistoric building techniques also underwent significant developments in the Neolithic Revolution. Nothing is known of the building ability of Palaeolithic peoples beyond what can be inferred from a few fragments of stone shelters, but in the New Stone Age some impressive structures were erected, primarily tombs and burial mounds and other religious edifices, but also, toward the end of the period, domestic housing in which sun-dried brick was first used.

In northern Europe, where the Neolithic transformation began later than around the eastern Mediterranean and lasted longer, huge stone monuments, of which Stonehenge in England is the outstanding example, still bear eloquent testimony to the technical skill, not to mention the imagination and mathematical competence, of the later Stone Age societies.

10.12 POLICE TECHNOLOGY MODUS OPERANDI

If a criminal has been successful using certain methods and techniques, he tends to repeat the same procedure. Thus, evidence of characteristic behaviour and procedure at the crime scene can serve to identify the offender if the pattern is recognized by the police.

Although classification of criminals according to operational methods and manners has lost a good deal of its effectiveness in the investigation of burglaries, large law-enforcement agencies still maintain modus operandi files to recognize patterns of behaviour, to associate a group of crimes with a single perpetrator, and to enable them to predict the next target of the criminal.

The modus operandi file is most effective in crimes involving personal contact, such as felonies against persons, confidence games, and forgery. Enhanced computer storage and search capabilities have rekindled interest in modus operandi methods of investigation, even in such cases as burglaries and robberies.

10.13 MODES OF TECHNOLOGICAL TRANSMISSION

Another aspect of the cumulative character of technology that will require further investigation is the manner of transmission of technological innovations. This is an elusive problem, and it is necessary to accept the phenomenon of simultaneous or parallel invention in cases in which there is insufficient evidence to show the transmission of ideas in one direction or another.

The mechanics of their transmission have been enormously improved in recent centuries by the printing press and other means of communication and also by the increased facility with which travellers visit the sources of innovation and carry ideas back to their own homes. Traditionally, however, the major mode of transmission has been the movement of artefacts and craftsmen.

Trade in artefacts has ensured their widespread distribution and encouraged imitation. Even more important, the migration of craftsmen--whether the itinerant metalworkers of early civilizations or the German rocket engineers whose expert knowledge was acquired by both the Soviet Union and the United States after World War II--has promoted the spread of new technologies.

The evidence for such processes of technological transmission is a reminder that the material for the study of the history of technology comes from a variety of sources. Much of it relies, like any historical examination, on documentary matter, although this is sparse for the early civilizations because of the general lack of interest in technology on the part of scribes and chroniclers.

For these societies, therefore, and for the many millennia of earlier unrecorded history in which slow but substantial technological advances were made, it is necessary to rely heavily upon archaeological evidence. Even in connection with the recent past, the historical understanding of the processes of rapid industrialization can be made deeper and more vivid by the study of "industrial archaeology."

Much valuable material of this nature has been accumulated in museums, and even more remains in the place of its use for the observation of the field worker. The historian of technology must be prepared to use all these sources, and to call upon the skills of the archaeologist, the engineer, the architect, and other specialists as appropriate.

10.14 TELETEXT AND VIDEOTEX

In addition to fax transmission, there exist two other forms of electronic still-image transmission that have been adopted in several parts of the world. These two forms of still-image transmission-- teletext and videotex--are usually employed to access information from computer databases.

In teletext, still images are transmitted in several scan lines of a television signal, which may be sent by either radio or cable. In the teletext receiver, the still images are captured line by line and stored in local memory for subsequent display on a television screen. In videotex, still images are transmitted digitally over the public switched telephone network by use of a modem.

At the receiver, the digital signal is recovered from a modem and is stored in local memory, again for subsequent display on a television screen. Videotex has been deployed in Germany, the United Kingdom, France, and Japan, while teletext has been deployed on several cable-television systems. The French videotex system, Teletel, has constructed a widespread packet-switched data network that supports several million Minitel videotex terminals.

11. IMPACTING CONSIDERATIONS

11.1 TECHNIQUES

Essentially, techniques are methods of creating new tools and products of tools, and the capacity for constructing such artefacts is a determining characteristic of manlike species. Other species make artifacts: bees build elaborate hives to deposit their honey, birds make nests, and beavers build dams. But these attributes are the result of patterns of instinctive behaviour and cannot be varied to suit rapidly changing circumstances.

Man, in contrast with other species, does not possess highly developed instinctive reactions but does have the capacity to think systematically and creatively about techniques. He can thus innovate and consciously modify his environment in a way no other species has achieved. An ape may on occasion use a stick to beat bananas from a tree: a man can fashion the stick into a cutting tool and remove a whole bunch of bananas.

Somewhere in the transition between the two, the hominid, or the first manlike species, emerges. By virtue of his nature as a toolmaker, man is therefore a technologist from the beginning, and the history of technology encompasses the whole evolution of man.

In using his rational faculties to devise techniques and modify his environment, man has attacked problems other than those of survival and the production of wealth with which the term technology is usually associated today.

The technique of language, for example, involves the manipulation of sounds and symbols in a meaningful way, and similarly the techniques of artistic and ritual creativity represent other aspects of the technological incentive.

This article does not deal with these cultural and religious techniques, but it is valuable to establish their relationship at the outset because the history of technology reveals a profound interaction between the incentives and opportunities of technological innovation on the one hand and the socio-cultural conditions of the human group within which they occur on the other.

11.2 SOUND RECORDIST

The main task of the recordist during live recording is to get "clean" dialogue that eliminates background noise and seems to correspond to

59

the space between speaker and camera. Most of the non-synchronous dialogue, sound effects, and music can be added and adjusted later. During shooting the sound recordist adjusts the sound by setting levels, altering microphone placement, and mixing (combining signals if there is more than one microphone).

Major technical and aesthetic reshaping is left for the postproduction phase when overhead is lower, the facilities are more sophisticated, and alternative versions can be created. It is also the job of the sound personnel to record wild sound (important sound effects and non-synchronous dialogue) and ambient sound (the inherent sound of the location). Ambient sound is added to the sound track during postproduction to maintain continuity between takes. Usually, wild sound and music are also adjusted and added then.

11.3 CAMERA SUPPORTS

The camera must be mounted on a substantial support to avoid extraneous movements while film is being exposed. In its simplest form this is a heavy tripod structure, with sturdy but smooth-moving adjustments and casters, so that the exact desired position can be quickly reached. Often a heavy dolly, holding both the camera and a seated cameraman, is used. This can be pushed or driven around the set. When shots from elevated positions are to be used, both camera and cameraman are carried on the end of a crane, also on a dolly. In some cases the assemblage is smoothly driven to follow the action being pictured, such as movement along a street. If the surface being traversed is not smooth, rails, resembling train tracks, must be laid on the floor or ground for the dolly. The camera may be freed from the tripod or dolly and carried by the operator by means of a body brace and gyroscope stabilizer. One such support is the Steadicam, which eliminates the tell-tale motions of the hand-held camera.

11.4 MAGOSIAN INDUSTRY

This is the stone-tool technology in which an advanced Levallois technique was employed for the production of flakes for the manufacture of other tools, together with a punch technique for the production of micro-lithic artefacts. Projectile points were produced by pressure flaking.

The site for which the industry is named is located in northern Uganda. Other sites in central and southern Africa that are dated to the Pleistocene epoch (which occurred from 1,600,000 to 10,000 years ago) are often considered to represent the same material culture and hunting-and-gathering adaptation.

11.5 IMPACT OF COMPUTER SYSTEMS

Many products used in everyday life now incorporate computer systems: programmable, computer-controlled VCRs in the living room, programmable microwave ovens in the kitchen, programmable thermostats to control heating and cooling systems--the list seems endless.

This section will survey a few of the major areas where computers currently have--or will likely soon have--a major impact on society. As noted below, computer technology not only has solved problems but also has created some, including a certain amount of culture shock as individuals attempt to deal with the new technology.

A major role of computer science has been to alleviate such problems, mainly by making computer systems cheaper, faster, more reliable, and easier to use.

11.6 COMPUTER SYSTEMS

The preceding sections give some idea of the pervasiveness of computer technology in society. Many products used in everyday life now incorporate computer systems: programmable, computer-controlled VCRs in the living room, programmable microwave ovens in the kitchen, programmable thermostats to control heating and cooling systems--the list seems endless.

This section will survey a few of the major areas where computers currently have--or will likely soon have--a major impact on society. As noted below, computer technology not only has solved problems but also has created some, including a certain amount of culture shock as individuals attempt to deal with the new technology.

A major role of computer science has been to alleviate such problems, mainly by making computer systems cheaper, faster, more reliable, and easier to use.

12. INNOVATION AND INDUSTRY

12.1 INNOVATION

The word innovation raises a problem of great importance in the history of technology. Strictly, an innovation is something entirely new, but there is no such thing as an unprecedented technological innovation because it is impossible for an inventor to work in a vacuum and, however ingenious his invention, it must arise out of his own previous experience.

The task of distinguishing an element of novelty in an invention remains a problem of patent law down to the present day, but the problem is made relatively easy by the possession of full documentary records covering previous inventions in many countries. For the millennium of the Middle Ages, however, few such records exist, and it is frequently difficult to explain how particular innovations were introduced to Western Europe.

The problem is especially perplexing because it is known that many inventions of the period had been developed independently and previously in other civilizations, and it is sometimes difficult if not impossible to know whether something is spontaneous innovation or an invention that had been transmitted by some as yet undiscovered route from those who had originated it in other societies.

The problem is important because it generates a conflict of interpretations about the transmission of technology. On the one hand there is the theory of the diffusionists, according to which all innovation has moved westward from the long-established civilizations of the ancient world, with Egypt and Mesopotamia as the two favourite candidates for the ultimate source of the process.

On the other hand is the theory of spontaneous innovation, according to which the primary determinant of technological innovation is social need. Scholarship is as yet unable to solve the problem so far as technological advances of the Middle Ages are concerned because much information is missing.

But it does seem likely that at least some of the key inventions of the period--the windmill and gunpowder are good examples--were developed spontaneously. It is quite certain, however, that others, such as silk working, were transmitted to the West, and, however original the contribution of Western civilization to technological innovation, there can be no doubt at all that in its early centuries at least it looked to the East for ideas and inspiration.

12.1 PRODUCTION

There have been technological innovations of great significance in many aspects of industrial production during the 20th century. It is worth observing, in the first place, that the basic matter of industrial organization has become one of self-conscious innovation, with organizations setting out to increase their productivity by improved techniques.

Methods of work study, first systematically examined in the United States at the end of the 19th century, were widely applied in U.S. and European industrial organizations in the first half of the 20th century, evolving rapidly into scientific management and the modern studies of industrial administration, organization and method, and particular managerial techniques.

The object of these exercises has been to make industry more efficient and thus to increase productivity and profits, and there can be no doubt that they have been remarkably successful, if not quite as successful as some of their advocates have maintained.

Without this superior industrial organization it would not have been possible to convert the comparatively small workshops of the 19th century into the giant engineering establishments of the 20th with their mass-production and assembly-line techniques.

The rationalization of production, so characteristic of industry in the 20th century, may thus be legitimately regarded as the result of the application of new techniques that form part of the history of technology since 1900.

12.3 TRANSPORT AND COMMUNICATIONS

Transport and communications provide an example of a revolution within the Industrial Revolution, so completely were the modes transformed in the period 1750-1900. The first improvements in Britain came in roads and canals in the second half of the 18th century.

Although of great economic importance, these were not of much significance in the history of technology, as good roads and canals had existed in continental Europe for at least a century before their adoption in Britain. A network of hard-surfaced roads was built in France in the 17th and early 18th centuries and copied in Germany.

Pierre Trésaguet of France improved road construction in the late 18th century by separating the hard-stone wearing surface from the rubble substrata and providing ample drainage. Nevertheless, by the beginning of the 19th century, British engineers were beginning to innovate in both road- and canal-building techniques, with J.L. McAdam's inexpensive and long-wearing road surface of compacted stones and Thomas Telford's well-engineered canals. The outstanding innovation in transport, however, was the application of steam power, which occurred in three forms.

12.4 ANIMATION

The basis of all animation is the building up, frame by frame, of the moving picture by exact timing and choreography of both movement and sound. All film movement is achieved by projecting during every second of time a certain number of frames, normally 24, each a still photograph minutely varied from its predecessor, which record the successive phases of the subject's movement before the camera.

The same motion, or a stylized or caricatured version of it, can be achieved by "stop-motion" or "stop-action" cinematography, the frame-by-frame photographing of a similarly phased series of drawings or the phased movement of such objects as puppets, marionettes, or commercial products. And, as in live filming, the camera itself can create movement by tracking into a scene or panning across it.

The great majority of animated films are short and have always been so for obvious reasons. When each second of action requires, for the fullest animation, 24 adjustments of the image, a minute's action may call for many hundreds of drawings.

The range of techniques in animation production is broad. The basic form is the simple, outlined figure, however, that moves against a simple, outlined background.

12.5 MUSIC IN MOTION PICTURES

There are two basic kinds of music; underscoring is usually background orchestration motivated by dramatic considerations, and source music is that which may be heard by the characters. Neither is likely to be recorded during shooting. Because a performance is usually divided into separate shots that take minutes or hours to prepare, it would be extremely difficult to produce a continuous musical performance.

Thus, most musical numbers are filmed to synchronize with a playback track. The songs and accompaniment are prerecorded, so that during filming the musician is mouthing the words or faking the playing in time to the track recorded earlier.

Whether music is chosen from music libraries or specially composed for the film, it cannot be prepared until the picture has been edited. The first step in scoring is spotting, or deciding which scenes shall have music and where it is to begin and end. The music editor then uses an editing console to break down each use of music, or cue, into fractions of seconds.

Recording is done on a recording stage, with individual musicians or groups of instruments miked individually and separated from one another, sometimes by acoustical partitions. In this case the conductor's function of balancing the instrumentalists may be left to the scoring mixer, who can adjust each track later.

12.6 MOTION PICTURES FOR SCIENTIFIC PURPOSES

As soon as motion pictures were invented, they were applied in the recording of scientific phenomena. The recording of an experiment in which a number of things happen at about the same time is especially appropriate for motion-picture recording.

12.7 TIME-LAPSE CINEMATOGRAPHY

There are many occasions on which the cinematography can be carried out at normal speeds. There are other situations, however, in which the changes occur very slowly, so slowly that the eye does not discern the change. One example is the opening of a flower blossom.

In such a situation the technique used is to take successive pictures at intervals of, for example, an hour, taking great care not to move the camera or the plant and to project the resulting film at normal motion-picture speed. The projected picture will disclose many details in the development from the bud to the completed flower that are not apparent in ordinary visual observation. Other phenomena can be studied in this way.

These techniques require merely a standard camera that can take single exposures, plus a timed triggering device that can take the exposures at the desired intervals. The rest of the technique is mostly a matter of preventing undesired motions of camera and subject.

12.8 DEVELOPMENT OF RADIO TECHNOLOGY

Early in the 19th century, Michael Faraday, an English physicist, demonstrated that an electric current can produce a local magnetic field and that the energy in this field will return to the circuit when the current is stopped or changed. James Clerk Maxwell, professor of experimental physics at Cambridge, in 1864 proved mathematically that any electrical disturbance could produce an effect at a considerable distance from the point at which it occurred and predicted that electromagnetic energy could travel outward from a source as waves moving at the speed of light.

12.9 HERTZ: RADIO-WAVE EXPERIMENTS

At the time of Maxwell's prediction there were no known means of propagating or detecting the presence of electromagnetic waves in space. It was not until about 1888 that Maxwell's theory was tested by Heinrich Hertz, who demonstrated that Maxwell's predictions were true at least over short distances by installing a spark gap (two conductors separated by a short gap) at the centre of a parabolic metal mirror.

A wire ring connected to another spark gap was placed about five feet (1.5 metres) away at the focus of another parabolic collector in line with the first. A spark jumping across the first gap caused a smaller spark to jump across the gap in the ring five feet away. Hertz showed that the waves travelled in straight lines and that they could be reflected by a metal sheet just as light waves are reflected by a mirror.

12.10 LIBRARY REFORMATTING

In response to this problem, libraries have developed several preservation strategies. The most important method of preserving library materials has been reformatting. Brittle and crumbling books and photographs are preserved by photographing them on microfilm or, in some cases, by using scanners to create digital images on magnetic or optical disc. These less vulnerable formats can then be preserved in archives.

Reformatting also enables the inclusion of library materials in other media, such as multimedia information services. The drawback of this process, of course, is the issue of technological obsolescence. If reformatting relies on technology that becomes obsolete, the preservation effort is seriously compromised. The task of reformatting all materials that used acidic paper, nitrate films, or other degradable materials is monumental, generally requiring cooperation between many libraries and a substantial infusion of government funds.

12.11 PUBLISHING

This is an account of the selection, preparation, and marketing of printed matter from its origins in ancient times to the present. The activity has grown from small beginnings into a vast and complex industry responsible for the dissemination of all manner of cultural material; its impact upon civilization is impossible to calculate.

This article treats the history and development of book, newspaper, and magazine publishing in its technical and commercial aspects. The preparation and dissemination of written communication is followed from its beginnings in the ancient world to the modern period. For additional information on the preparation of early manuscripts, see writing.

A more detailed examination of printing technology can be found in printing. The dissemination of published material via electronic media is treated in information processing. For a discussion of reference-book publishing, see the articles encyclopaedia; dictionary.

12.12 ELECTRONIC BANKING

The banking business has been revolutionized by computer technology. Deposits and withdrawals are instantly logged into a customer's account, which is perhaps stored on a remote computer. Computer-generated monthly statements are unlikely to contain any errors unless they arise during manual entry of check amounts.

The technology of electronic funds transfer, supported by computer networking, allows the amount of a grocery bill to be immediately deducted from the customer's bank account and transferred to that of the grocery store.

Similarly, networking allows individuals to obtain cash instantly and almost worldwide by simply stepping up to an automated teller machine (ATM) and providing the proper card and personal identification number (popularly known as a PIN).

The downside of this technology is the potential for security problems. Intruders can see packets travelling on a network (e.g., being transported via a satellite link) and can perhaps interpret them (if not carefully encrypted) to obtain confidential information on financial transactions.

Network access to personal accounts has the potential to let intruders not only see how much money an individual has but also to transfer some of it elsewhere.

12.13 NON-AEROSPACE PRODUCTS

Non-aerospace products and systems, general-usage offshoots of aerospace technology, constitute a fourth area of industry workload, not significant as a proportion of total workload except in the United States.

Examples of new applications of aerospace-developed technology include many medical aids, ranging from a thimble-sized cardiac-pacemaker power source to a computerized hospital-management system; in transportation, hydrofoil ships and boats, vastly improved railroad cars, truck and bus transmissions, motor-vehicle traffic-monitoring systems, inflatable crash-bag systems for automobiles, and turbine propulsion systems for both land and water vehicles; in construction, prefabricated low-cost housing and lightweight, super-strong structural materials; in electric power, a number of types of nuclear and non-nuclear power-generating systems; and, in environmental control, a broad range of antipollution systems.

12.14 EDUCATION AND SOCIAL SERVICES

Of all Manchester's pioneer cultural achievements, none has prospered more than the Victoria University of Manchester. After its foundation in 1851 at a site in Quay Street, the college received a charter in 1872 and began growth on its present site in 1873. By 1880 it had combined with member colleges in both Leeds and Liverpool to form a federal institution. Since becoming a separate body again in 1903, the university has grown to become one of the largest in Britain.

The faculty of technology has become autonomous as an Institute of Science and Technology, and, with the establishment of the University of Salford in 1967 and the growth of a large polytechnic, there are now four institutions of higher learning in and near the city.

The city provides the complete range of social and welfare services within the British system, but its special strength lies in health services and medical education. The Victoria University of Manchester has the largest medical school in Western Europe; it is linked to three large

68

groups of teaching hospitals that provide specialist treatment. One of the most distinguished of these is the Christie Hospital, a major centre for cancer research.

12.15 CONSULTING

The rapid proliferation of new discoveries, products, and markets in the electrical and electronics industries has made it difficult for workers in the field to maintain the range of skills required to manage their activities. Consulting engineers, specializing in new fields, are employed to study and recommend courses of action.

The educational background required for these functions tends to be highest in basic and applied research. In most major laboratories a doctorate in science or engineering is required to fill leadership roles. Most positions in design, product development, and supervision of manufacture and quality control require a master's degree.

In the high-technology industries typical of modern electronics, an engineering background at not less than the bachelor's level is required to assess competitive factors in sales engineering to guide marketing strategy.

69

13. INSTRUMENTS AND ENGINEERING

13.1 AEROSPACE ENGINEERING

This is also called **AERONAUTICAL ENGINEERING,** or **ASTRONAUTICAL ENGINEERING,** field of engineering concerned with the design, development, construction, testing, and operation of vehicles operating in the Earth's atmosphere or in outer space. In 1958 the first definition of aerospace engineering appeared, considering the Earth's atmosphere and the space above it as a single realm for development of flight vehicles. Today the more encompassing aerospace definition has commonly replaced the terms aeronautical engineering and astro-nautical engineering.

The design of a flight vehicle demands a knowledge of many engineering disciplines. It is rare that one person takes on the entire task; instead, most companies have design teams specialized in the sciences of aerodynamics, propulsion systems, structural design, materials, avionics, and stability and control systems. No single design can optimize all of these sciences, but rather there exist compromised designs that incorporate the vehicle specifications, available technology, and economic feasibility.

13.2 AUTOMATIC INSTRUMENTS

Water power, clockwork, steam, and electricity have all been used at various times to "power" musical instruments, enabling them to produce sound automatically. Examples include church bells, automatic organs, musical clocks, automatic pianos and harpsichords, music boxes, calliopes, and even automatic orchestras. Most of the impetus behind this phenomenon ceased with the development of the phonograph and other recording devices of the 20th century.

13.3 ELECTRIC AND ELECTRONIC INSTRUMENTS

The development of electricity led not only to its use for mechanical purposes--for example, to control the key action and wind flow in the organ--but also as a means of amplification (*e.g.,* in the vibraphone). With advances in electronics technology, players can now also make use of computers to generate and store tones and musical patterns.

The growth of companies manufacturing electronic and digital instruments has been rapid, and the use of electronic equipment such as sound synthesizers and tape recorders to produce and combine sound unrelated to the musical scale has become common.

13.4 INTERNAL-COMBUSTION ENGINE

Electricity does not constitute a prime mover, for however important it may be as a form of energy it has to be derived from a mechanical generator powered by water, steam, or internal combustion. The internal-combustion engine is a prime mover, and it emerged in the 19th century as a result both of greater scientific understanding of the principles of thermodynamics and of a search by engineers for a substitute for steam power in certain circumstances.

In an internal-combustion engine the fuel is burned in the engine: the cannon provided an early model of a single-stroke engine; and several persons had experimented with gunpowder as a means of driving a piston in a cylinder. The major problem was that of finding a suitable fuel, and the secondary problem was that of igniting the fuel in an enclosed space to produce an action that could be easily and quickly repeated.

The first problem was solved in the mid-19th century by the introduction of town gas supplies, but the second problem proved more intractable as it was difficult to maintain ignition evenly. The first successful gas engine was made by Étienne Lenoir in Paris in 1859. It was modelled closely on a horizontal steam engine, with an explosive mixture of gas and air ignited by an electric spark on alternate sides of the piston when it was in mid-stroke position.

Although technically satisfactory, the engine was expensive to operate, and it was not until the refinement introduced by the German inventor Nikolaus Otto in 1878 that the gas engine became a commercial success. Otto adopted the four-stroke cycle of induction-compression-firing-exhaust that has been known by his name ever since. Gas engines became extensively used for small industrial establishments, which could thus dispense with the upkeep of a boiler necessary in any steam plant, however small.

13.5 FUEL AND POWER

There were no fundamental innovations in fuel and power before the breakthrough of 1945, but there were several significant developments in techniques that had originated in the previous century. An outstanding development of this type was the internal-combustion engine, which was continuously improved to meet the needs of road vehicles and airplanes.

The high-compression engine burning heavy-oil fuels, invented by Rudolf Diesel in the 1890s, was developed to serve as a submarine power unit in World War I and was subsequently adapted to heavy road haulage duties and to agricultural tractors. Moreover, the sort of development that had transformed the reciprocating steam engine into the steam turbine occurred with the internal-combustion engine, the gas turbine replacing the reciprocating engine for specialized purposes such as aero-engines, in which a high power-to-weight ratio is important.

Admittedly, this adaptation had not proceeded very far by 1945, although the first jet-powered aircraft were in service by the end of the war. The theory of the gas turbine, however, had been understood since the 1920s at least, and in 1929 Sir Frank Whittle, then taking a flying instructor's course with the Royal Air Force, combined it with the principle of jet propulsion in the engine for which he took out a patent in the following year.

But the construction of a satisfactory gas-turbine engine was delayed for a decade by the lack of resources, and particularly by the need to develop new metal alloys that could withstand the high temperatures generated in the engine. This problem was solved by the development of a nickel-chromium alloy, and with the gradual solution of the other problems work went on in both Germany and Britain to seize a military advantage by applying the jet engine to combat aircraft.

13.6 ELECTRONIC MATERIALS

Between 1955 and 1990, improvements and innovations in semiconductor technology increased the performance and decreased the cost of electronic materials and devices by a factor of one million-- an achievement unparalleled in the history of any technology. Along with this extraordinary explosion of technology has come an exponentially upward spiral of the capital investment necessary for manufacturing operations. In order to maintain cost-effectiveness and flexibility, radical changes in materials and manufacturing operations will be necessary.

14. INTERGRATION OF ELECTRONICS AND ENGINEERING

14.1 SILICON

Bulk semiconductor silicon for the manufacture of integrated circuits (sometimes referred to as electronic-grade silicon) is the purest material ever made commercially in large quantities. One of the most important factors in preparing this material is control of such impurities as boron, phosphorus, and carbon (not to be confused with the dopants added later during circuit production). For the ultimate levels of integrated-circuit design, stray contaminant atoms must constitute less than 0.1 part per trillion of the material.

For fabrication into integrated circuits, bulk semiconductor silicon must be in the form of a single-crystal material with high crystalline perfection and the desired charge-carrier concentration. The size of the silicon ingot, or boule, has been scaled up in recent years, in order to provide wafers of increasing diameter that are demanded by the economics of integrated-circuit manufacturing. Most commonly, a 60-kilogram (130-pound) charge is grown to an ingot with a diameter of 200 millimetres (8 inches), but the semiconductor industry will soon require ingots as large as 300 millimetres. The ingots are then converted into wafers by machining and chemical processes.

14.2 WEATHER

In the latter part of the 1800s during the homesteading of the Great Plains, farmers were for the most part successful because the weather was generally favourable. Severe drought conditions in the 1890s, however, shattered the illusion that life on the plains was necessarily good and drove out many settlers. This pattern of boom-and-bust farming recurred several times: good weather and high-production years were followed by periods of drought, economic ruin, and serious soil erosion. The worst drought and resultant soil degradation occurred during the 1930s in the area of the Dust Bowl. Average wheat and corn yields fell by as much as 75 percent. Worse, millions of tons of valuable topsoil were lost.

As a result of this climate-induced disaster, the federal government established the U.S. Soil Conservation Service to help farmers protect the soil. Decades later new crop strains better adapted to regional climates were developed, and irrigation and chemical fertilizers were made available to take advantage of the new genetic strains and to promote production. These advances, together with the development of

73

such technological aids as tractors, mechanical harvesters, irrigation pumps, and agrochemicals (*e.g.,* pesticides and herbicides), have increased the productivity (average yield of grain per harvested area) of the Great Plains by 200 to 300 percent since the 1930s. Total production (productivity multiplied by total harvested area) also has risen. In addition, the amount of year-to-year variability in yield, relative to the average yield, for most grains has decreased over time.

Yet, while the relative variability (year-to-year variability as a percentage of long-term average yield) has generally declined for crops in recent years, the absolute variability (year-to-year variability in yield by itself) has increased on the whole in spite of all the technological advances.

This has led to an ongoing debate about the relative role of climate versus the role of technological advances in influencing both average-yield and yield-variability trends during the 1960s and '70s. While agrotechnology appears to have been the prime factor behind the general increase in annual grain yield, there is some doubt as to its actual contribution to the favourable decrease of variability in relative yield.

Some investigators, most notably the American geographer Richard Warrick, have argued that this decrease cannot be entirely attributed to modern farming practices. Only if the weather anomalies of the 1960s and '70s had been as bad as those from roughly 1930 to the late 1950s could agrotechnology be credited with the reduced impact of climatic stress on crop yields.

Warrick and his associates compiled an index of the severity of summer droughts in the Great Plains in the period 1931-77, which indicates quite clearly that since the late 1950s weather conditions have not been bad enough to test the hypothesis that technological advances have truly reduced the annual variations of grain yield. Given this situation, it is risky to assume that modern technology in the "breadbasket" of North America can stably maintain yearly productivity.

14.3 STRUCTURAL

Structural integrants of garden and landscape include structures closely related to the earth, enclosure structures, shelter structures, engineering structures, and special buildings.

Earth-related structures include paving (walks, roads, terraces, patios) and change-of-level structures (retaining walls, steps, ramps, bridges), which must be made of materials that will resist decay, such as brick, stone, concrete, asphalt. These structures provide the connections for movement and circulation and the areas for intensive gathering, social use, or active recreation. They embody a complex technology.

14.4 ENCLOSURES

Enclosure structures, such as walls and fences, are designed to control vision or movement or both. They may be of various heights, three to ten feet (one to three metres) or more, and of many materials: brick, stone, or concrete masonry; wood; metal; sheet materials such as glass, plastic, asbestos, pressed boards. Because they are at eye level and extend and connect buildings, they are very important in intimate visual design.

14.5 END OF TELEGRAPH ERA

After World War II much new technology became available that radically changed the telegraph industry. Old wire lines were too expensive to maintain and were replaced by coaxial cable and microwave links. Very wide-bandwidth channels became available, allowing transmission speeds limited only by the capabilities of the terminal equipment. These new transmission media were later augmented by satellite links and fibre optic transmission lines. In 1974 the Westar satellite, providing enormous capacity for all types of telecommunication, was placed in operation by Western Union.

These new transmission channels were complemented by new electronic technology including transistors, integrated circuits, and various microelectronics devices that reduced costs and improved performance. With the advent of the digital computer in the 1960s the trend toward entirely digital communication began.

The facsimile telegraph was perfected in the 1930s and was widely used for sending photographs and other graphic information over telephone and telegraph lines in an analogue transmission system. By the 1980s, however, analogue facsimile was virtually replaced by the digital fax machine. In many offices, fax machines have replaced other types of communication, including telegrams, TWX, Telex, and, in many cases, the postal service.

In the face of changing technology, the Western Union Telegraph Company was reorganized as the Western Union Corporation in 1988 to handle money transfers and related services. It sold its international private line service to Tele-Columbus AG of Switzerland, the Westar satellite was sold to GM Hughes Electronics Corporation of the United States, and AT&T acquired Western Union's business services group. The telegraph, which started in 1837, had been replaced in most applications in developed countries by digital data transmission systems based on computer technology.

14.6 STEAM LOCOMOTIVE

First was the evolution of the railroad: the combination of the steam locomotive and a permanent travel way of metal rails. Experiments in this conjunction in the first quarter of the 19th century culminated in the Stockton & Darlington Railway, opened in 1825, and a further five years of experience with steam locomotives led to the Liverpool and Manchester Railway, which, when it opened in 1830, constituted the first fully timetabled railway service with scheduled freight and passenger traffic relying entirely on the steam locomotive for traction.

This railway was designed by George Stephenson, and the locomotives were the work of Stephenson and his son Robert, the first locomotive being the famous *Rocket*, which won a competition held by the proprietors of the railway at Rainhill, outside Liverpool, in 1829. The opening of the Liverpool and Manchester line may fairly be regarded as the inauguration of the Railway Era, which continued until World War I. During this time railways were built across all the countries and continents of the world, opening up vast areas to the markets of industrial society.

Locomotives increased rapidly in size and power, but the essential principles remained the same as those established by the Stephensons in the early 1830s: horizontal cylinders mounted beneath a multi-tubular boiler with a firebox at the rear and tender carrying supplies of water and fuel. This was the form developed from the *Rocket*, which had diagonal cylinders, being itself a stage in the transition from the vertical cylinders, often encased by the boiler, which had been typical of the earliest locomotives (except Trevithick's Penydarren engine, which had a horizontal cylinder).

Meanwhile, the construction of the permanent way underwent a corresponding improvement on that which had been common on the preceding tramroads: wrought-iron, and eventually steel, rails replaced the cast-iron rails, which cracked easily under a steam locomotive, and well-aligned track with easy gradients and substantial supporting civil-engineering works became a commonplace of the railroads of the world.

14.7 INFORMATION PROCESSING PRINTERS

Computer printers are commonly divided into two general classes according to the way they produce images on paper: impact and non-impact. In the first type, images are formed by the print mechanism making contact with the paper through an ink-coated ribbon. The mechanism consists either of print hammers shaped like characters or of a print head containing a row of pins that produce a pattern of dots in the form of characters or other images.

Most non-impact printers form images from a matrix of dots, but they employ different techniques for transferring images to paper. The most popular type, the laser printer, uses a beam of laser light and a system of optical components to etch images on a photoconductor drum from which they are carried via electrostatic photocopying to paper. Light-emitting diode (LED) printers resemble laser printers in operation but direct light from energized diodes rather than a laser onto a photoconductive surface.

Ion-deposition printers make use of technology similar to that of photocopiers for producing electrostatic images. Another type of non-impact printer, the ink-jet printer, sprays electrically charged drops of ink onto the print surface.

14.8 DIESEL ENGINE

One objectionable feature of the full diesel was the necessity of a high-pressure, injection air compressor. Not only was energy required to drive the air compressor, but the sudden expansion of the air compressed to 6.9 mega-pascals when it entered the cylinder in which the pressure was only about 3.4 to 4.1 mega-pascals resulted in a refrigerating effect that delayed ignition. Diesel had needed high-pressure air with which to introduce powdered coal into the cylinder; when liquid petroleum replaced powdered coal as fuel, a pump could be made to take the place of the high-pressure air compressor.

There were a number of ways in which a pump could be used. In England the Vickers Company used what was called the common-rail method, in which a battery of pumps maintained the fuel under pressure in a pipe running the length of the engine with leads to each cylinder. From this rail (or pipe) fuel-supply line, a series of injection valves admitted the fuel charge to each cylinder at the right point in its cycle. Another method employed cam-operated jerk, or plunger-type, pumps, to deliver fuel under momentarily high pressure to the injection valve of each cylinder at the right time.

The elimination of the injection air compressor was a step in the right direction, but there was yet another problem to be solved: the engine exhaust contained an excessive amount of smoke, even at outputs well within the horsepower rating of the engine and even though there was enough air in the cylinder to burn the fuel charge without leaving a discoloured exhaust that normally indicated overload.

Engineers finally realized that the problem was that the momentarily high-pressure injection air exploding into the engine cylinder had diffused the fuel charge more efficiently than the substitute mechanical fuel nozzles were able to do, with the result that without the air compressor, the fuel had to search out the oxygen atoms to complete the combustion process, and since oxygen makes up only 20 percent of the air, each atom of fuel had only one chance in five of encountering an atom of oxygen. The result was improper burning of the fuel.

The usual design of a fuel-injection nozzle introduced the fuel into the cylinder in the form of a cone spray, with the vapour radiating from the nozzle, rather than in a stream or jet. Very little could be done to diffuse the fuel more thoroughly. Improved mixing had to be accomplished by imparting additional motion to the air, most commonly by induction-produced air swirls or a radial movement of the air, called squish, or both, from the outer edge of the piston toward the centre.

Various methods have been employed to create this swirl and squish. Best results are apparently obtained when the air swirl bears a definite relation to the fuel-injection rate. Efficient utilization of the air within the cylinder demands a rotational velocity that causes the entrapped air to move continuously from one spray to the next during the injection period, without extreme subsidence between cycles.

14.9 AGRICULTURAL METHODS

Fertilizers may be added to soil in solid, liquid, or gaseous forms, the choice depending on many factors. Generally, the farmer tries to obtain satisfactory yield at minimum cost in money and labour.

Manure can be applied as a liquid or a solid. When accumulated as a liquid from livestock areas, it may be stored in tanks until needed and then pumped into a distributing machine or into a sprinkler irrigation system. The method reduces labour, but the noxious odours are objectionable. The solid-manure spreader conveys the material to the field, shreds it, and spreads it uniformly over the land. The process can be carried out during convenient times, including winter, but rarely when the crop is growing.

Application of granulated or pelleted solid fertilizer has been aided by improved equipment design. Such devices, depending on design, can deposit fertilizer at the time of planting, side-dress a growing crop, or broadcast the material. Fertilizer attachments are available for most tractor-mounted planters and cultivators and for grain drills and some types of ploughs. They deposit fertilizer with the seed when planted, without damage to the seed, yet the nutrient is readily available during early growth. Placement of the fertilizer varies according to the types of crops; some crops require banding above the seed, while others are more successful when the fertilizer band is below the seed.

The use of liquid and ammonia fertilizers is growing, particularly of anhydrous ammonia, which is handled as a liquid under pressure but changes to gas when released to atmospheric pressure. Anhydrous ammonia, however, is highly corrosive, inflammable, and rather dangerous if not handled properly; thus, application equipment is quite specialized.

Typically, the applicator is a chisel-shaped blade with a pipe mounted on its rear side to conduct the ammonia five to six inches (13 to 15 centimetres) below the surface. Pipes are fed from a pressure tank mounted above. Mixed liquid fertilizers containing nitrogen, phosphorus, and potassium may be applied directly to the surface--by field sprayers where close-growing crops are raised. Large areas can be covered rapidly by use of aircraft, which can distribute both liquid and dry fertilizer.

14.10 ELECTRICITY

All the principles of generating electricity had been worked out in the 19th century, but by its end these had only just begun to produce electricity on a large scale. The 20th century has witnessed a colossal expansion of electrical power generation and distribution. The general pattern has been toward ever-larger units of production, using steam from coal- or oil-fired boilers.

Economies of scale and the greater physical efficiency achieved as higher steam temperatures and pressures were attained both reinforced this tendency. U.S. experience indicates the trend: in the first decade of the century a generating unit with a capacity of 25,000 kilowatts with pressures up to 200-300 pounds per square inch at 400°-

79

500° F (about 200°-265° C) was considered large, but by 1930 the largest unit was 208,000 kilowatts, with pressures of 1,200 pounds per square inch at a temperature of 725° F, while the amount of fuel necessary to produce a kilowatt-hour of electricity and the price to the consumer had fallen dramatically.

As the market for electricity increased, so did the distance over which it was transmitted, and the efficiency of transmission required higher and higher voltages. The small direct-current generators of early urban power systems were abandoned in favour of alternating-current systems, which could be adapted more readily to high voltages. Transmission over a line of 155 miles (250 kilometres) was established in California in 1908 at 110,000 volts; Hoover Dam in the 1930s used a line of 300 miles (480 kilometres) at 287,000 volts.

The latter case may serve as a reminder that hydroelectric power, using a fall of water to drive water turbines, has been developed to generate electricity where the climate and topography make it possible to combine production with convenient transmission to a market. Remarkable levels of efficiency have been achieved in modern plants. One important consequence of the ever-expanding consumption of electricity in the industrialized countries has been the linking of local systems to provide vast power grids, or pools, within which power can be shifted easily to meet changing local needs for current.

14.11 PLASTICS

The quality of plasticity is one that had been used to great effect in the crafts of metallurgy and ceramics. The use of the word plastics as a collective noun, however, refers not so much to the traditional materials employed in these crafts as to new substances produced by chemical reactions and moulded or pressed to take a permanent rigid shape.

The first such material to be manufactured was Parkesine, developed by the British inventor Alexander Parkes. Parkesine, made from a mixture of chloroform and castor oil, was "a substance hard as horn, but as flexible as leather, capable of being cast or stamped, painted, dyed or carved ... "

The words are from a guide to the International Exhibition of 1862 in London, at which Parkesine won a bronze medal for its inventor. It was soon followed by other plastics, but apart from celluloid, a cellulose nitrate composition using camphor as a solvent and produced in solid form (as imitation horn for billiard balls) and in sheets (for

80

men's collars and photographic film), these had little commercial success until the 20th century.

The early plastics had relied upon the large molecules in cellulose, usually derived from wood pulp. Leo H. Baekeland, a Belgian-U.S. inventor, introduced a new class of large molecules when he took out his patent for Bakelite in 1909. Bakelite is made by the reaction between formaldehyde and phenolic materials at high temperatures; the substance is hard, infusible, and chemically resistant (the type known as thermosetting plastic). As a non-conductor of electricity it proved to be exceptionally useful for all sorts of electrical appliances.

The success of Bakelite gave a great impetus to the plastics industry, to the study of coal-tar derivatives and other hydrocarbon compounds, and to the theoretical understanding of the structure of complex molecules. This activity led to new dyestuffs and detergents, but it also led to the successful manipulation of molecules to produce materials with particular qualities such as hardness or flexibility.

Techniques were devised, often requiring catalysts and elaborate equipment, to secure these polymers--that is, complex molecules produced by the aggregation of simpler structures. Linear polymers give strong fibres, film-forming polymers have been useful in paints, and mass polymers have formed solid plastics.

14.12 FUSE/FUZE EXPLOSIVES TECHNOLOGY

This is the device used for firing explosives in blasting operations, in fireworks, and in military projectiles.

The blasting safety fuse, employed to fire an explosive from a distance or after a delay, is a hollow cord filled with a mixture resembling black powder and designed to propagate burning at a slow and steady rate. The far end of the fuse is usually embedded in the explosive charge. Detonating cord, also called Cordeau and Primacord, is a hollow cord filled with an explosive material. It is fired by a detonator and is capable of initiating the detonation of certain other explosives at any number of points and in any desired pattern.

The United States and some other military forces have adopted the "z" spelling for the device in ordnance munitions; the fuze sets off the munition, regulates its functioning, and causes it to perform only under predetermined conditions. It is distinct from the primer or firing pin that initiates the launching of a rocket or artillery shell. Impact fuzes function as they hit the target.

Time fuzes delay the functioning for a certain period from the starting time. Command fuzes function on signal from a remote-control point. Proximity fuzes function when the munitions carrying them approach to within a given distance of the target. Inferential fuzes infer that a target is nearby if certain conditions are present.

15. AUTOMATION AND COMPLICATED DEVICES

15.1 ENVIRONMENTAL WORKS

This is the infrastructure that provides cities and towns with water supply, waste disposal, and pollution control services. They include extensive networks of reservoirs, pipelines, treatment systems, pumping stations, and waste disposal facilities. These municipal works serve two important purposes: they protect human health and safeguard environmental quality.

Treatment of drinking water helps to prevent the spread of waterborne diseases such as cholera, dysentery, and typhoid fever, and proper waste treatment and disposal practices prevent degradation of ecosystems and neighbourhoods. Similarly, cleaning the air of pollutant gases and particles as they are generated prevents adverse effects on both human health and the environment.

Steady population growth, urbanization, and industrial development place steadily increasing demands on existing infrastructure, and these demands in turn create a need for the planning, design, and construction of new environmental works. Because the provision, operation, and maintenance of these works require a major investment of public funds, concerned citizens as well as municipal officials and decision makers should be familiar with the basic concepts of environmental works technology.

This article presents an introduction to the fundamentals of environmental works. Its main focus is on the modern facilities and systems that provide communities with water, dispose of waste, and prevent pollution.

15.2 SPACE SUITS

The designing of a much more complicated device, such as a space suit, presents more intricate problems. A space suit is a complete miniature world, a self-contained environment that must supply everything needed for an astronaut's life, as well as comfort. The suit must provide a pressurized interior, without which an astronaut's blood would boil in the vacuum of space.

The consequent pressure differential between the inside and the outside of the suit is so great that when inflated the suit becomes a distended, rigid, and unyielding capsule. Special joints were designed to give the astronaut as much free movement as possible. The best

83

engineering has not been able to provide as much flexibility of movement as is desirable; to compensate for that lack, attention has been directed toward the human-factors design of the tools and devices that an astronaut must use.

In addition to overcoming pressurization and movement problems, a space suit must provide oxygen; a system for removing excess products of respiration, carbon dioxide and water vapour; protection against extreme heat, cold, and radiation; protection for the eyes in an environment in which there is no atmosphere to absorb the sun's rays; facilities for speech communication; and facilities for the temporary storage of body wastes.

This is such an imposing list of human requirements that an entire technology has been developed to deal with them and, indeed, with the provision of simulated environments and procedures for testing and evaluating space suits.

15.3 MECHANICAL ENGINEERING

Closely linked with the iron and steel industry was the rise of mechanical engineering, brought about by the demand for steam engines and other large machines, and taking shape for the first time in the Soho workshop of Boulton and Watt in Birmingham, where the skills of the precision engineer, developed in manufacturing scientific instruments and small arms, were first applied to the construction of large industrial machinery.

The engineering workshops that matured in the 19th century played a vital part in the increasing mechanization of industry and transport. Not only did they deliver the looms, locomotives, and other hardware in steadily growing quantities, but they also transformed the machine tools on which these machines were made. The lathe became an all-metal, power-driven machine with a completely rigid base and a slide rest to hold the cutting tool, capable of more sustained and vastly more accurate work than the hand- or foot-operated wooden-framed lathes that preceded it.

Drilling and slotting machines, milling and planing machines, and a steam hammer invented by James Nasmyth (an inverted vertical steam engine with the hammer on the lower end of the piston rod), were among the machines devised or improved from earlier woodworking models by the new mechanical engineering industry. After the middle of the 19th century, specialization within the machinery industry became more pronounced, as some manufacturers concentrated on vehicle production while others devoted themselves to the particular needs of industries such as coal mining, papermaking, and sugar refining.

This movement toward greater specialization was accelerated by the establishment of mechanical engineering in the other industrial nations, especially in Germany, where electrical engineering and other new skills made rapid progress, and in the United States, where labour shortages encouraged the development of standardization and mass-production techniques in fields as widely separated as agricultural machinery, small arms, typewriters, and sewing machines. Even before the coming of the bicycle, the automobile, and the airplane, therefore, the pattern of the modern engineering industry had been clearly established.

The dramatic increases in engineering precision, represented by the machine designed by British mechanical engineer Sir Joseph Whitworth in 1856 for measuring to an accuracy of 0.000001 inch (even though such refinement was not necessary in everyday workshop practice), and the corresponding increase in the productive capacity of the engineering industry, acted as a continuing encouragement to further mechanical innovation.

15.4 COMPUTERS IN THE WORKPLACE

Computers are omnipresent in the workplace. Word processors--computer software packages that simplify the creation and modification of textual documents--have largely replaced the typewriter. Electronic mail has made it easy to transmit textual messages (possibly containing embedded picture and sound files) worldwide, using computers, cellular telephones, and specially equipped televisions via telephone, satellite, and cable television networks.

Office automation has become the term for linking workstations, printers, database systems, and other tools by means of a local area network (LAN). An eventual goal of office automation has been termed the "paperless office." Although such changes ultimately make office work much more efficient, they have not been without cost in purchasing and frequently upgrading the necessary hardware and software and in training workers to use the new technology.

Computer-integrated manufacturing (CIM) is a relatively new technology arising from the application of many computer science sub-disciplines to support the manufacturing enterprise. The technology of

85

CIM emphasizes that all aspects of manufacturing should be not only computerized as much as possible but also linked into an integrated whole via a computer communication network. For example, the design engineer's workstation should be linked into the overall system so that design specifications and manufacturing instructions may be sent automatically to the shop floor.

The inventory databases should be linked in as well, so product inventories may be incremented automatically and supply inventories decremented as manufacturing proceeds. An automated inspection system (or a manual inspection station supplied with online terminal entry) should be linked to a quality-control system that maintains a database of quality information and alerts the manager if quality is deteriorating and possibly even provides a diagnosis as to the source of any problems that arise.

Automatically tracking the flow of products from station to station on the factory floor allows an analysis program to identify bottlenecks and recommend replacement of faulty equipment. In short, CIM has the potential to enable manufacturers to build cheaper, higher quality products and thus improve their competitiveness. Implementing CIM is initially costly, of course, and progress in carrying out this technology has been slowed not only by its cost but also by the lack of standardized interfaces between the various CIM components and by the slow acceptance of standardized communication protocols to support integration.

Although the ideal of CIM is perhaps just beyond reach at the present time, manufacturers are now able to improve their operations by, for example, linking robot controllers to mainframes for easy and correct downloading of revised robot instructions. Also available are elaborate software packages that simplify the building of databases for such applications as inventories, personnel statistics, and quality control and that incorporate tools for data analysis and decision support.

15.5 LIGHTHOUSE AUTOMATION

The acetylene-gas illumination system, being fully automatic and reliable, enabled automatic lights to be operated early on. Its main use today is in buoys, which inherently have to operate unattended. Automation on a large scale, bringing considerable savings in operating costs, came after the advent of electrical equipment and technology and the demise of compressed-air fog signals.

Unattended lights are now designed to be automatic and self-sustaining, with backup plant brought on-line automatically upon failure of any component of the system. The status of the station is monitored from a remote control centre via landline, radio, or satellite link. Power is provided from public electricity supplies (where practicable), with backup provided by diesel generators or storage batteries.

Where solar power with storage batteries is used, the batteries must have sufficient capacity to operate the light during the hours of darkness. In tropical and subtropical regions, day and night are of approximately equal duration throughout the year, but in temperate and Polar Regions the days become longer and the nights shorter during the summer, and vice versa in winter.

In these areas, solar power has to operate on an annual "balance sheet" basis, with excess charge being generated and stored in large batteries during the summer so that a reserve can be drawn upon in winter. Canada and Norway successfully operate solar-powered lights of this type in their Arctic regions.

16. NEW TECHNIQUES OF MANUFACTURE

16.1 COMMODITIES

Following the dramatic expansion of the European nations into the Indian Ocean region and the New World, the commodities of these parts of the world found their way back into Europe in increasing volume. These commodities created new social habits and fashions and called for new techniques of manufacture.

Tea became an important trade commodity but was soon surpassed in volume and importance by the products of specially designed plantations, such as sugar, tobacco, cotton, and cocoa. Sugar refining, depending on the crystallization of sugar from the syrupy molasses derived from the cane, became an important industry.

So did the processing of tobacco, for smoking in clay pipes (produced in bulk at Delft and elsewhere) or for taking as snuff. Cotton had been known before as an Eastern plant, but its successful transplantation to the New World made much greater quantities available and stimulated the emergence of an important new textile industry.

The woollen cloth industry in Britain provided a model and precedent upon which the new cotton industry could build. Already in the Middle Ages, the processes of cloth manufacture had been partially mechanized upon the introduction of fulling mills and the use of spinning wheels.

But in the 18th century the industry remained almost entirely a domestic or cottage one, with most of the processing being performed in the homes of the workers, using comparatively simple tools that could be operated by hand or foot. The most complicated apparatus was the loom, but this could usually be worked by a single weaver, although wider cloths required an assistant. It was a general practice to install the loom in an upstairs room with a long window giving maximum natural light.

Weaving was regarded as a man's work, spinning being assigned to the women of the family (hence, "spinsters"). The weaver could use the yarn provided by up to a dozen spinsters, and the balanced division of labour was preserved by the weaver's assuming responsibility for supervising the cloth through the other processes, such as fulling. Pressures to increase the productivity of various operations had already produced some technical innovations by the first half of the 18th century.

The first attempts at devising a spinning machine, however, were not successful; and without this, John Kay's technically successful flying shuttle (a device for hitting the shuttle from one side of the loom to the other, dispensing with the need to pass it through by hand) did not fulfil an obvious need. It was not until the rapid rise of the cotton cloth industry that the old, balanced industrial system was seriously upset and that a new, mechanized system, organized on the basis of factory production, began to emerge.

16.2 DIAGNOSIS, DNA PROBES

Karyo-typing requires a great deal of time and effort and may not always provide conclusive information. It is most useful in identifying very large defects involving hundreds or even thousands of genes.

Newer techniques such as fluorescent in situ hybridization (FISH) have much higher rates of sensitivity and specificity. FISH also provides results more quickly because no cell culture is required. This technique can detect smaller genetic deletions involving one to five genes. It is also useful in detecting moderate-sized deletions such as those causing Prader-Willi syndrome, which is characterized by a rounded face, low forehead, and mental retardation.

The analysis of individual genes has been greatly enhanced by the development of recombinant DNA technology. Small DNA fragments can be isolated, and unlimited amounts of cloned material can be produced. Once cloned, the various genes and gene products can be used to study gene function in healthy individuals and those with disease. Recombinant DNA methods can detect any change in DNA, down to a one-base-pair change out of the three billion base pairs in the genome.

DNA probes are labelled with radioactive isotopes or fluorescent dyes and used to identify persons who are carriers for autosomal recessive conditions. Disorders that can be detected using this technique include haemophilia A, polycystic kidney disease, sickle-cell disease, Huntington's chorea, cystic fibrosis, and haemochromatosis.

16.3 DEVELOPMENTS IN WATER TREATMENT

In addition to quantity of supply, water quality is also of concern. Even the ancients had an appreciation for the importance of water purity. Sanskrit writings from as early as 2000 BC tell how to purify foul water by boiling and filtering. But it was not until the middle of the 19th century that a direct link between polluted water and disease

89

(cholera) was proved. And it was not until the end of that same century that the German bacteriologist Robert Koch proved the germ theory of disease, establishing a scientific basis for the treatment and sanitation of drinking water.

Water treatment is the alteration of a water source in order to achieve a quality that meets specified goals. At the end of the 19th century and the beginning of the 20th, the main goal was elimination of deadly waterborne diseases. The treatment of public drinking water to remove pathogenic, or disease-causing, micro-organisms began about that time. Treatment methods included sand filtration as well as the use of chlorine for disinfection. The virtual elimination of diseases such as cholera and typhoid in developed countries proved the success of this water treatment technology.

In developing countries, waterborne disease is still the principal water quality concern. In industrialized nations, however, concern has shifted to the chronic health effects related to chemical contamination. For example, trace amounts of certain synthetic organic substances in drinking water are suspected of causing cancer in humans. The added goal of reducing such health risks is seen in the continually increasing number of factors included in drinking-water standards.

6.4 CONSERVATION

This is the planned management of a natural resource or the total environment of a particular ecosystem to prevent exploitation, pollution, destruction, or neglect and to ensure the future use of the resource.

Although the idea of conservation is probably as old as the human species, the use of the word in its present context is relatively recent. Over the years conservation has acquired many connotations: to some it has meant the protection of wild nature, to others the sustained production of useful materials from the resources of the Earth.

The most widely accepted definition, presented in 1980 in *World Conservation Strategy* by the International Union for Conservation of Nature and Natural Resources, is that of "the management of human use of the biosphere so that it may yield the greatest sustainable benefit while maintaining its potential to meet the needs and aspirations of future generations." The document defines the objectives of the conservation of living resources as: maintenance of essential ecological processes and life-support systems, preservation of genetic diversity, and guarantee of the sustainable use of species and ecosystems.

90

More generally, conservation involves practices that perpetuate the resources of the Earth on which human beings depend and that maintain the diversity of living organisms that share the planet. This includes such activities as the protection and restoration of endangered species, the careful use or recycling of scarce mineral resources, the rational use of energy resources, and the sustainable use of soils and living resources.

Conservation is necessarily based on a knowledge of ecology, the science concerned with the relationship between life and the environment, but ecology itself is based on a wide variety of disciplines, and conservation involves human feelings, beliefs, and attitudes as well as science and technology.

16.5 ELECTRIC CONNECTIONS

The performance of today's electronic systems (and photonic systems as well) is limited significantly by interconnection technology, in which components and subsystems are linked by conductors and connectors. Currently, very fine gold or copper wiring, as thin as 30 micro-metres, is used to carry electric current to and from the many pads along the sides or ends of a microchip to other components on a circuit board.

The capacitance involved in such circuitry slows down the flow of electrons and, hence, of information. However, by integrating several chips into a single multi-chip module, in which the chips are connected on a shared substrate by various conducting materials (such as metalized film), the speed of information flow, can be increased, thus improving the assembly's performance.

Ideally, all the chips in a single module would be fabricated simultaneously on the same wafer, but in practice this is not feasible: Silicon crystal manufacture is still subject to an average of one flaw per wafer, meaning that at least one of the many chips cut from each wafer is scrapped. If the whole wafer area were dedicated to a single multifunction assembly, that one flaw would scrap the entire module.

Multi-chip modules are therefore made up of as many as five microchips bonded to a silicon or ceramic substrate on which resistors and capacitors have been constructed with thin films.

Typical materials used in a multi-chip module include the substrate; gold paste conductors applied in an additive process resembling silk screen printing; vitreous glazes to insulate the gold paste conductors from subsequent film layers; a series of thin films made with tantalum

Andreas Sofroniou

nitride, titanium, palladium, and plated gold; and a final package of silicone rubber.

16.6 ART AND TECHNOLOGY RELATIONSHIP

To what extent the arts can influence the sciences is still uncertain. If "visual thinking" precedes conceptual thought, or if there is a basic background element common to both visual and conceptual thought, changes in perceptions that art either reflects or stimulates may generate responses in conceptual theorizing and, in this manner, affect developments in science as well as in philosophy. But it is on social and psychological theory and on the perceived shape of history that developments in the arts may be expected to have the strongest effect.

The association of artists with technology, especially in the visual arts and architecture, has been far closer and more durable than it has been with science, and the requirements and achievements of artists have often led to technological discoveries. Literary artists, on the other hand, have, particularly in the periods of great scientific discovery, such as the 17th century, been influenced more by science than by technology.

It might seem that the new technologies of the mass media of communication, by permitting more pervasive dissemination of particular works of art throughout society, are likely to increase the impact of art on people's personalities and modes of existence. But the mass media may also have the opposite effect: by making of art a commonplace occurrence that is taken for granted but does not generate an enduring emotional response, by assimilating it more completely to "entertainment," the mass media may decrease the effects of works of art on people and, indeed, on the artists themselves and on their aesthetic experiences.

At the same time, the media may be increasing the psychological impact of non-artistic events, particularly those that resist being assimilated to entertainment, such as wars in a foreign country, which were easier to take for granted before the electronic media became fully developed. This has led to tendencies to substitute, especially in the visual arts, the structures of events or of social systems for aesthetic structures (for example, an art exhibition consisting of photographs of slum buildings and landlords' names).

While it is debatable whether the mass media have enhanced the impact of art, they have certainly increased the social status of the performing artists (and tend to transform all artists whom they capture into performers).

17. IMPACT OF TECHNOLOGICAL DEVELOPMENTS

17.1 TELESCOPE

Besides the telescope itself, the electronic computer has become the astronomer's most important tool. Indeed, the computer has revolutionized the use of the telescope to the point where the collection of observational data is now completely automated. The astronomer need only identify the object to be observed, and the rest is carried out by the computer and auxiliary electronic equipment.

A telescope can be set to observe automatically by means of electronic sensors appropriately placed on the telescope axis. Precise quartz or atomic clocks send signals to the computer, which in turn activates the telescope sensors to collect data at the proper time. The computer not only makes possible more efficient use of telescope time but also permits a more detailed analysis of the data collected than could have been done manually. Data analysis that would have taken a lifetime or longer to complete with a mechanical calculator can now be done within hours or even minutes with a high-speed computer.

Improved means of recording and storing computer data also have contributed to astronomical research. Optical disc data storage technology, such as the CD-ROM (compact disc read-only memory) or the WORM (write-once read-many) disc has provided astronomers with the ability to store and retrieve vast amounts of telescopic and other astronomical data. A 12-centimetre CD-ROM, for example, may hold up to 600 megabytes of data--the equivalent of 20 nine-track magnetic tapes or 1,500 floppy discs. A 13-centimetre WORM disc typically holds about 300 to 400 megabytes of data.

17.2 PUBLIC UTILITY

This is the enterprise that provides certain classes of services to the public, including common carrier transportation (buses, airlines, railroads, motor freight carriers, pipelines, etc.); telephone and telegraph; power, heat, and light; and community facilities for water, sanitation, and similar services. In most countries such enterprises are state-owned and state-operated, but in the United States they are mainly privately owned and are operated under close governmental regulation.

The classic explanation for the need to regulate public utilities is that they are enterprises in which the technology of production, transmission, and distribution almost inevitably leads to complete or partial monopoly--that they are, in a phrase, natural monopolies. The monopolistic tendency arises from economies of scale in the particular industry, from the large capital costs typical of such enterprises, from the inelasticity of demand among consumers of the service, from considerations of the excess capacity necessary to meet demand peaks, and other considerations. It is often also the case that the existence of competing parallel systems--of local telephones or natural gas, for example--would be inordinately expensive, wasteful, and inconvenient. Given the tendency to monopoly and the potential therefore of monopolistic pricing practices, public regulation has for more than a century been applied to certain classes of business.

In practice, regulation aims to ensure that the utility serves all who apply for and are willing and able to pay for its services, that it operates in a safe and adequate manner, that it serves all customers on equal terms, and that its rates are just and reasonable. All states have regulatory commissions, and the federal government has several, including the Interstate Commerce Commission, the Civil Aeronautics Board, the Federal Power Commission, the Federal Communications Commission, and the Securities and Exchange Commission.

17.3 HARVESTING AND CROP PROCESSING

Harvesting machinery is generally classified by crop: reapers for cutting cereal grains and threshers for separating the seed from the plant. The more modern combine cuts, threshes, and cleans the grain in one operation.

Corn (maize) harvesting is performed by mechanical corn pickers that snap the ears from the stalk so that only the grain and cobs are harvested. Corn shelling may be done mechanically in the field, after or with picking. Stripper-type cotton harvesters, which strip the entire plant of both open and unopened bolls, work best late in the season after frost has killed the green vegetative growth. Hay and forage machines include mowers, crushers, windrowers, field choppers, balers, and some machines that press the hay into wafers or pellets.

Grass, legumes, corn (maize), and other crops are often put into silos to keep them in a succulent and fermented state rather than stored dry as hay. To make silage, the crops must be cut up to permit tight packing in the silo, producing anaerobic fermentation and preventing formation of mould. Almost all silage crops are cut in the field with a forage harvester that cuts and chops the crop immediately or picks up and chops a windrow that has been cut and raked earlier.

Root crops are harvested with diggers and digger-pickers, which often pull up clods, stones, and vines with the crop. Though some machines carry workers who manually sort out extraneous material, this task is increasingly being performed mechanically. Modern sugar-beet harvesters lift the whole root from the ground, clean the earth from it, and deliver it to a bin or wagon. Sometimes the beet tops are removed before harvest of the roots and used for cattle feed. Peanuts (groundnuts) are lifted, vines and all, and allowed to dry before removal of the pods.

Tobacco-harvesting aids may be classified in three principal ways, according to the harvesting and curing methods used, which depend on the type of tobacco and its use. Flue-cured tobacco, a large plant that may stand three to four feet (90 to 120 centimetres) high, is harvested with machines that carry several workers who ride the lower platforms of the machines, cut the leaves, and place them on conveyor belts, where the leaves are tied mechanically or by hand.

Burley tobacco has usually been harvested by workers using a machete-type knife. After cutting, the large end of the stalk is fixed onto the sharpened end of a stick, which--when loaded with a number of stalks--is hung by hand in a tobacco barn for curing. Researchers are attempting to mechanize the cutting, impaling, and hanging of burley tobacco. Little has been done, however, toward the mechanization of the harvesting of the small aromatic tobacco leaves, which are grown in the shade, picked by hand, tied with a string, then hung for curing.

Tree-crop harvesting is accomplished by hand or with mechanical shakers. Vegetable crops such as asparagus, lettuce, and cabbage are still harvested largely by hand, though scarcity and high cost of field labour has led to some mechanization in this area, notably with tomatoes.

17.4 DOMESTIC ARCHITECTURE

Domestic architecture is produced for the social unit: the individual, family, or clan and their dependents, human and animal. It provides shelter and security for the basic physical functions of life and at times also for commercial, industrial, or agricultural activities that involve the family unit rather than the community. The basic requirements of domestic architecture are simple: a place to sleep, prepare food, eat, and perhaps work; a place that has some light and is protected from the weather. A single room with sturdy walls and roof, a door, a window, and a hearth are the necessities; all else is luxury.

17.5 VERNACULAR ARCHITECTURE

In much of the world today, even where institutions have been in a continuous process of change, dwelling types of ancient or prehistoric origin are in use. In the industrialized United States, for instance, barns are being built according to a design employed in Europe in the 1st millennium BC. The forces that produce a dynamic evolution of architectural style in communal building are usually inactive in the home and farm.

The life of the average person may be unaltered by the most fundamental changes in his institutions. He can be successively a slave, the subject of a monarchy, and a voting citizen, without having the means or the desire to change his customs, techniques, or surroundings. Economic pressure is the major factor that causes the average individual to restrict his demands to a level far below that which the technology of his time is capable of maintaining.

Frequently he builds new structures with old techniques because experiment and innovation are more costly than repetition. But in wealthy cultures economy permits and customs encourage architecture to provide conveniences such as sanitation, lighting, and heating, as well as separate areas for distinct functions, and these may come to be regarded as necessities. The same causes tend to replace the conservatism of the home with the aspirations of institutional architecture and to emphasize the expressive as well as the utilitarian function.

17.6 RETAIL

Computer technology has had a significant impact on retail stores. All but the smallest shops have replaced the old-fashioned cash register with a terminal linked to a computer system.

Some terminals require that the clerk type in the code for the item, but most checkout counters include a bar-code scanner, a device that directly reads into the computer the Universal Product Code (UPC) printed on each package. Cash-register receipts can then include brief descriptions of the items purchased (by fetching them from the computer database), and the purchase information is also relayed back to the computer to cause an immediate adjustment in the inventory data.

The inventory system can easily alert the manager when the supply of some item drops below a specified threshold. In the case of retail

96

chains linked by networks, the order for a new supply of an item may be automatically generated and sent electronically to the supply warehouse. In a less extensively automated arrangement, the manager can send in the order electronically by a dial-up link to the supplier's computer.

These developments have made shopping much more convenient. The checkout process is faster, checkout lines are shorter, and the desired item is more likely to be in stock. In addition, cash-register receipts contain much more information than a simple list of item prices; many receipts now include discount coupons based on the specific items purchased by the shopper.

If there is a downside, it is the need for shoppers to adjust psychologically to not seeing prices on the packages and to the feeling that perhaps the computer is overcharging (as indeed can happen when advertised sale prices are somehow not entered into the system).

Since the mid-1990s one of the most rapidly growing retail sectors, known as electronic commerce, or e-commerce, involves the use of the Internet and proprietary networks to facilitate business-to-business, consumer, and auction sales of everything imaginable--from computers and electronics to books, recordings, automobiles, and real estate.

17.7 MAGNETIC RESONANCE IMAGING

Magnetic resonance imaging (MRI) relies on the response of magnetic fields to short bursts of radio-frequency waves to produce computer images that provide structural and biochemical information about tissue. The process uses radio waves and is thus much safer than imaging using X rays or gamma rays.

This totally non-invasive but very expensive procedure is particularly useful in detecting cerebral oedema, abnormalities of the spine, and early-stage cancer. In examining the brain, spinal cord, urinary bladder, pelvic organs, and cancellous bone, MRI is the superior imaging technique. Because patients must lie quietly inside a narrow tube, MRI may raise anxiety levels in the patients, especially those with claustrophobia.

Another disadvantage of MRI is that it has a longer scanning time than CT, which makes it more sensitive to motion artefacts and thus of less value in scanning the chest or abdomen. Because of the strong magnetic field, MRI cannot be used if a pacemaker is present or if metal is present in critical areas such as the eye or brain.

MRI has largely supplanted arthrography, the injection of dye into a joint to visualize cartilage or ligament damage to the knee or shoulder, and myelography, the injection of dye into the spinal canal to visualize spinal cord or intervertebral disk abnormalities.

Multiple sclerosis, a disease with multiple foci of demyelination (loss of the myelin sheath of a nerve) in the brain, sometimes can be diagnosed using MRI. However, because the test is not sufficiently sensitive, a normal MRI cannot exclude the diagnosis.

Magnetic resonance angiography, a unique form of MRI technology, can be used to produce an image of flowing blood. This permits the visualization of arteries and veins without the need for needles, catheters, or contrast agents.

CT and MRI provide two-dimensional views of cross sections of the body and these images must be viewed in sequence by the radiologist. Computer technology now makes it possible to construct holograms that provide three-dimensional images from digital data obtained by conventional CT or MRI scanners. These holograms can be useful in locating lesions more precisely and in mapping the exact location of coronary arteries when planning bypass surgery or angioplasty.

17.8 ROBOT MANIPULATOR

The most widely accepted definition of an industrial robot is one developed by the Robotic Industries Association:

An industrial robot is a reprogrammable, multifunctional manipulator designed to move materials, parts, tools, or specialized devices through variable programmed motions for the performance of a variety of tasks.

The technology of robotics is concerned with the design of the mechanical manipulator and the computer systems used to control it. It is also concerned with the industrial applications of robots, which are described below.

The mechanical manipulator of an industrial robot is made up of a sequence of link and joint combinations. The links are the rigid members connecting the joints. The joints (also called axes) are the movable components of the robot that cause relative motion between adjacent links. There are five principal types of mechanical joints used to construct the manipulator.

Two of the joints are linear, in which the relative motion between adjacent links is translational, and three are rotary types, in which the relative motion involves rotation between links.

The manipulator can be divided into two sections:

(1) An arm-and-body, which usually consists of three joints connected by large links, and

(2) A wrist, consisting of two or three compact joints.

Attached to the wrist is a gripper to grasp a work part or a tool (e.g., a spot-welding gun) to perform a process. The two manipulator sections have different functions: the arm-and-body is used to move and position parts or tools in the robot's work space, while the wrist is used to orient the parts or tools at the work location. The arm-and-body section of most commercial robots is based on one of four configurations. Each of the anatomies, as they are sometimes called, provides a different work envelope (i.e., the space that can be reached by the robot's arm) and is suited to different types of applications.

17.9 DEVELOPMENTS IN SEWAGE TREATMENT

It used to be said that "the solution to pollution is dilution." When small amounts of sewage are discharged into a flowing body of water, a natural process of stream self-purification occurs. Densely populated communities generate such large quantities of sewage, however, that dilution alone does not prevent pollution. This makes it necessary to treat or purify wastewater to some degree before disposal.

The construction of centralized sewage treatment or water-pollution control plants began in the late 19th and early 20th centuries, principally in the United Kingdom and the United States. Instead of discharging sewage directly into a nearby body of water, it was first passed through a combination of physical, biological, and chemical processes that removed some or most of the pollutants. Also beginning in the 1900s, new sewage-collection systems were designed to separate storm water from domestic wastewater, so that treatment plants did not become overloaded during periods of wet weather.

After the middle of the 20th century, increasing public concern for environmental quality led to broader and more stringent regulation of wastewater disposal practices. Higher levels of treatment were required. For example, pre-treatment of industrial wastewater, with the aim of preventing toxic chemicals from interfering with the biological processes used at sewage treatment plants, often became a necessity. In fact, wastewater treatment technology advanced to the

99

point where it became possible to remove virtually all pollutants from sewage. This was so expensive, however, that such high levels of treatment were not usually justified.

Modern water-pollution control plants became large, complex facilities that required considerable amounts of energy for their operation. After the rise of oil prices in the 1970s, concern for energy conservation became a more important factor in the design of new pollution control systems. Consequently, land disposal and subsurface disposal of sewage began to receive increased attention where feasible. Such low-tech pollution control methods not only would help to conserve energy but also would serve to recycle nutrients and replenish groundwater supplies.

18. INDUSTRIALIZATION AND COMPUTING

18.1 INDUSTRIAL DOMINANCE

Industrialisation is the process of converting to a socioeconomic order in which industry is dominant.

How or why some agrarian societies have evolved into industrial states is not always fully understood. What is certainly known, though, is that the changes that took place in Britain during the Industrial Revolution of the late 18th and 19th centuries provided a prototype for the early industrializing nations of western Europe and North America.

Along with its technological components (*e.g.,* the mechanization of labour and the reliance upon inanimate sources of energy), the process of industrialization entailed profound social developments. The freeing of the labourer from feudal and customary obligations created a free market in labour, with a pivotal role for a specific social type, the entrepreneur. Cities drew large numbers of people off the land, massing workers in the new industrial towns and factories.

Later industrializers attempted to manipulate some of these elements. The Soviet Union, for instance, industrialized largely on the basis of forced labour and eliminated the entrepreneur, while in Japan strong state involvement stimulated and sustained the entrepreneur's role. Other states, notably Denmark and New Zealand, industrialized primarily by commercializing and mechanizing agriculture.

Although urban-industrial life offers unprecedented opportunities for individual mobility and personal freedom, it can exact high social and psychological tolls. Such various observers as Karl Marx and Émile Durkheim cited the "alienation" and "anomie" of individual workers faced by seemingly meaningless tasks and rapidly altering goals. The fragmentation of the extended family and community tended to isolate individuals and to countervail traditional values.

By the very mechanism of growth, industrialism appears to create a new strain of poverty, whose victims for a variety of reasons are unable to compete according to the rules of the industrial order. In the major industrialized nations of the late 20th century, such developments as automated technology, an expanding service sector, and increasing suburbanization signalled what some observers called the emergence of a post-industrial society.

18.2 COMPUTERS USES IN DESIGNING

101

Since the mid-1960s, computer technology has been continually developed to the point at which aircraft and engine designs can be simulated and tested in myriad variations under a full spectrum of environmental conditions prior to construction. As a result, practical consideration may be given to a series of aircraft configurations, which, while occasionally and usually unsuccessfully attempted in the past, can now be used in production aircraft.

These include forward swept wings, canard surfaces, blended body and wings, and the refinement of specialized airfoils (wing, propeller, and turbine blade). With this goes a far more comprehensive understanding of structural requirements, so that adequate strength can be maintained even as reductions are made in weight.

Complementing and enhancing the results of the use of computers in design is the pervasive use of computers on board the aircraft itself. Computers are used to test and calibrate the aircraft's equipment, so that, both before and during flight, potential problems can be anticipated and corrected.

Whereas the first autopilots were devices that simply maintained an aircraft in straight and level flight, modern computers permit an autopilot system to guide an aircraft from takeoff to landing, incorporating continuous adjustment for wind and weather conditions and ensuring that fuel consumption is minimized.

In the most advanced instances, the role of the pilot has been changed from that of an individual who continuously controlled the aircraft in every phase of flight to a systems manager who oversees and directs the human and mechanical resources in the cockpit.

The use of computers for design and in-flight control is synergistic, for more radical designs can be created when there are on-board computers to continuously adapt the controls to flight conditions.

The degree of inherent stability formerly desired in an aircraft design called for the wing, fuselage, and empennage (tail assembly) of what came to be conventional size and configurations, with their inherent weight and drag penalties. By using computers that can sense changes in flight conditions and make corrections hundreds and even thousands of times a second--far faster and more accurately than any pilot's capability--aircraft can be deliberately designed to be unstable.

Wings can, if desired, be given a forward sweep, and tail surfaces can be reduced in size to an absolute minimum (or, in a flying wing layout, eliminated completely). Airfoils can be customized not only for a

102

particular aircraft's wing or propeller but also for particular points on those components.

18.3 ELECTRONIC ENCYCLOPAEDIAS

Given the rapid pace of technological advancement in the contemporary world, it was to be expected that encyclopaedia publishers would seek ways to exploit new technologies in the field of information storage, retrieval, and distribution. During the 1960s and '70s these new technologies revolutionized the manner in which article text was generated, modified as needed, and composed and output for printing.

The computer terminal, typically linked to a large mainframe computer where the encyclopaedia's contents were stored as an electronic database on magnetic tape or disc, became the key to editorial production. By the 1980s and '90s the phenomenal growth of telecommunications networks and personal computer systems presented a new possibility to the publishing industry--the delivery of encyclopaedic databases through a medium other than the printed page.

Many general and specialized encyclopaedias now publish electronic versions of their print sets, either as CD-ROM (compact disc read-only memory) products or as online services. As computer technology continues to develop and is used with greater sophistication, there exists the further possibility that the electronic encyclopaedia will become less a version of the print set than a product in its own right, presenting the database in a manner best suited to exploit the advantages of the electronic medium.

One advantage of the electronic medium is the huge storage capacity that it offers at very low cost. Freed from the expense of printing more pages and binding more volumes, electronic encyclopaedias are able to offer many more articles than their print versions. These articles are also more accessible: in addition to the alphabetical indexes compiled by editors for the print sets, many electronic encyclopaedias feature high-speed search software that can retrieve an exhaustive set of files from their databases in response to specific queries posed by readers.

The most obvious advantage of electronic encyclopaedias is in their "multimedia" capabilities, with animated graphics, recorded sound, and video recordings supplementing the text, photographs, and line drawings inherited from the print medium. With the development of more sophisticated data-processing applications, there arises the potential for truly "interactive" encyclopaedias, which would allow readers to retrieve, manipulate, and classify information according to their own designs.

18.4 FLIGHT SIMULATOR

The flight simulator is an electronic or mechanical system for training airplane and spacecraft pilots and crew members by simulating flight conditions. The purpose of simulation is not to completely substitute for actual flight training but to thoroughly familiarize students with the vehicle concerned before they undergo expensive and possibly dangerous actual flight training. Simulation also is useful for review and for familiarizing pilots with new modifications to existing craft.

Two early flight simulators appeared in England within a decade after the first flight of Orville and Wilbur Wright. They were designed to enable pilots to simulate simple aircraft manoeuvres in three dimensions: nose up or down; left wing high and right low, or vice versa; and yawing to left or right. It took until 1929, however, for a truly effective simulator, the Link Trainer, to appear, devised by Edwin A. Link, a self-educated aviator and inventor from Binghamton, New York. By then, airplane instrumentation had been developed sufficiently to permit "blind" flying on instruments alone, but training pilots to do so involved considerable risk.

Link built a model of an airplane cockpit equipped with instrument panel and controls that could realistically simulate all the movements of an airplane. Pilots could use the device for instrument training, manipulating the controls on the basis of instrument readings so as to maintain straight and level flight or controlled climb or descent with no visual reference to any horizon except for the artificial one on the instrument panel. The trainer was modified as aircraft technology advanced. Commercial airlines began to use the Link Trainer for pilot training, and the U.S. government began purchasing them in 1934, acquiring thousands more as World War II approached.

Technological advances during the war, particularly in electronics, helped to make the flight simulator increasingly realistic. The use of efficient analogue computers in the early 1950s led to further improvements. Airplane cockpits, controls, and instrument displays had by then become so individualized that it was no longer feasible to use a generalized trainer to prepare pilots to fly anything but the simplest light planes. By the 1950s, the U.S. Air Force was using simulators that precisely replicated the cockpits of its planes.

During the early 1960s electronic digital and hybrid computers were adopted, and their speed and flexibility revolutionized simulation systems. Further advances in computer and programming technology, notably the development of virtual-reality simulation, have made it possible to reproduce highly complex real-life conditions.

18.5 SUPERCOMPUTER

A supercomputer belongs to a class of extremely powerful computers. The term is commonly applied to the fastest high-performance systems available at any given time. Such computers are used primarily for scientific and engineering work requiring exceedingly high-speed computations.

Supercomputers have certain distinguishing features. Unlike conventional computers, they usually have more than one CPU (central processing unit), which contains circuits for interpreting program instructions and executing arithmetic and logic operations in proper sequence).

The use of several CPUs to achieve high computational rates is necessitated by the physical limits of circuit technology. Electronic signals cannot travel faster than the speed of light, which thus constitutes a fundamental speed limit for signal transmission and circuit switching.

This limit has almost been reached owing to miniaturization of circuit components, dramatic reduction in the length of wires connecting circuit boards, and innovation in cooling techniques (e.g., in various supercomputer systems, processor and memory circuits are immersed in a cryogenic fluid to achieve the low temperatures at which they operate fastest).

Rapid retrieval of stored data and instructions is required to support the extremely high computational speed of CPUs. Therefore, most supercomputers have a very large storage capacity, as well as a very fast input/output capability.

Still another distinguishing characteristic of supercomputers is their use of vector arithmetic--i.e., they are able to operate on pairs of lists of numbers rather than on mere pairs of numbers.

For example, a typical supercomputer can multiply a list of hourly wage rates for a group of factory workers by a list of hours worked by members of that group to produce a list of dollars earned by each worker in roughly the same time that it takes a regular computer to calculate the amount earned by just one worker.

Andreas Sofroniou

Supercomputers were originally used in applications related to national security, including nuclear-weapons design and cryptography. Today, they are also routinely employed by the aerospace, petroleum, and automotive industries.

In addition, supercomputers have found wide application in areas involving engineering or scientific research, as, for example, in studies of the structure of subatomic particles and of the origin and nature of the universe. Supercomputers have also become an indispensable tool in weather forecasting: predictions are now based on numerical models.

19. RENAISSANCE AND SUBSEQUENT SCIENCES

19.1 RENAISSANCE

According to medieval scientists, matter was composed of four elements--earth, air, fire, and water--whose combinations and permutations made up the world of visible objects. The cosmos was a series of concentric spheres in motion, the farther ones carrying the stars around in their daily courses.

At the centre was the globe of Earth, heavy and static. Motion was either perfectly circular, as in the heavens, or irregular or naturally downward, as on Earth. The Earth had three landmasses--Europe, Asia, and Africa--and was unknown and uninhabitable in its southern zones. Human beings, the object of all creation, were composed of four humours--black and yellow bile, blood, and phlegm--and the body's health was determined by the relative proportions of each.

The cosmos was alive with a universal consciousness with which people could interact in various ways, and the heavenly bodies were generally believed to influence human character and events, although theologians worried about free will.

These views were an amalgam of classical and Christian thought and, from what can be inferred from written sources, shaped the way educated people experienced and interpreted phenomena. What people who did not read or write books understood about nature is more difficult to tell, except that belief in magic, good and evil spirits, witchcraft, and forecasting the future was universal.

The church might prefer that Christians seek their well-being through faith, the sacraments, and the intercession of Mary and the saints, but distinctions between acceptable and unacceptable belief in hidden powers were difficult to make or to maintain. Most clergy shared the common beliefs in occult forces and lent their authority to them.

The collaboration of formal doctrine and popular belief had some of its most terrible consequences during the Renaissance, such as pogroms against Jews and witch-hunts, in which the church provided the doctrines of Satanic conspiracy and the inquisitorial agents and popular prejudice supplied the victims, predominantly women and marginal people.

Among the formally educated, if not among the general population, traditional science was transformed by the new heliocentric, mechanistic, and mathematical conceptions of Copernicus, Harvey, Kepler, Galileo, and Newton. Historians of science are increasingly

107

reluctant to describe these changes as a revolution, since this implies too sudden and complete an overthrow of the earlier model.

Aristotle's authority gave way very slowly and only the first of the great scientists mentioned above did his work in the period under consideration. Still, the Renaissance made some important contributions toward the process of paradigm shift, as the 20th-century historian of science Thomas Kuhn called major innovations in science.

Humanist scholarship provided both originals and translations of ancient Greek scientific works--which enormously increased the fund of knowledge in physics, astronomy, medicine, botany, and other disciplines--and presented as well alternative theories to those of Ptolemy and Aristotle. Thus, the revival of ancient science brought heliocentric astronomy to the fore again after almost two millennia.

Renaissance philosophers, most notably Jacopo Zabarella, analyzed and formulated the rules of the deductive and inductive methods by which scientists worked, while certain ancient philosophies enriched the ways in which scientists conceived of phenomena. Pythagoreanism, for example, conveyed a vision of a harmonious geometric universe that helped form the mind of Copernicus.

In mathematics the Renaissance made its greatest contribution to the rise of modern science. Humanists included arithmetic and geometry in the liberal arts curriculum; artists furthered the geometrization of space in their work on perspective; Leonardo da Vinci perceived, however faintly, that the world was ruled by "number."

The interest in algebra in the Renaissance universities, according to the 20th-century historian of science George Sarton, "was creating a kind of fever." It produced some mathematical theorists of the first rank, including Niccolò Tartaglia and Girolamo Cardano. If they had done nothing else, Renaissance scholars would have made a great contribution to mathematics by translating and publishing, in 1544, some previously unknown works of Archimedes, perhaps the most important of the ancients in this field.

If the Renaissance role in the rise of modern science was more that of midwife than of parent, in the realm of technology the proper image is the Renaissance magus, manipulator of the hidden forces of nature. Working with medieval perceptions of natural processes, engineers and technicians of the 15th and 16th centuries achieved remarkable results and pushed the traditional cosmology to the limit of its explanatory powers.

108

This may have had more to do with changing social needs than with changes in scientific theory. Warfare was one catalyst of practical change that stimulated new theoretical questions. With the spread of the use of artillery, for example, questions about the motion of bodies in space became more insistent and mathematical calculation more critical.

The manufacture of guns also stimulated metallurgy and fortification; town planning and reforms in the standards of measurement were related to problems of geometry. The Renaissance preoccupation with alchemy, the parent of chemistry, was certainly stimulated by the shortage of precious metals, made more acute by the expansion of government and expenditures on war.

The most important technological advance of all, because it underlay progress in so many other fields, strictly speaking, had little to do with nature. This was the development of printing, with movable metal type, about the mid-15th century in Germany. Johannes Gutenberg is usually called its inventor, but in fact many people and many steps were involved. Block printing on wood came to the West from China between 1250 and 1350, papermaking came from China by way of the Arabs to 12th-century Spain, whereas the Flemish technique of oil painting was the origin of the new printers' ink. Three men of Mainz-- Gutenberg and his contemporaries Johann Fust and Peter Schöffer-- seem to have taken the final steps, casting metal type and locking it into a wooden press. The invention spread like the wind, reaching Italy by 1467, Hungary and Poland in the 1470s, and Scandinavia by 1483. By 1500 the presses of Europe had produced some six million books.

Without the printing press it is impossible to conceive that the Reformation would have ever been more than a monkish quarrel or that the rise of a new science, which was a cooperative effort of an international community, would have occurred at all. In short, the development of printing amounted to a communications revolution of the order of the invention of writing; and, like that prehistoric discovery, it transformed the conditions of life.

The communications revolution immeasurably enhanced human opportunities for enlightenment and pleasure on one hand and created previously undreamed-of possibilities for manipulation and control on the other. The consideration of such contradictory effects may guard us against a ready acceptance of triumphalist conceptions of the Renaissance or of historical change in general.

19.2 DEVELOPMENT OF ELECTROMAGNETISM

Electromagnetic technology began with Faraday's discovery of induction in 1831 (see above). His demonstration that a changing magnetic field induces an electric current in a nearby circuit showed that mechanical energy can be converted to electric energy. It provided the foundation for electric power generation, leading directly to the invention of the dynamo and the electric motor. Faraday's finding also proved crucial for lighting and heating systems.

The early electric industry was dominated by the problem of generating electricity on a large scale. Within a year of Faraday's discovery, a small hand-turned generator in which a magnet revolved around coils was demonstrated in Paris. In 1833 there appeared an English model that featured the modern arrangement of rotating the coils in the field of a fixed magnet.

By 1850 generators were manufactured commercially in several countries. Permanent magnets were used to produce the magnetic field in generators until the principle of the self-excited generator was discovered in 1866. (A self-excited generator has stronger magnetic fields because it uses electromagnets powered by the generator itself.) In 1870 Zénobe Théophile Gramme, a Belgian manufacturer, built the first practical generator capable of producing a continuous current. It was soon found that the magnetic field is more effective if the coil windings are embedded in slots in the rotating iron armature.

The slotted armature, still in use today, was invented in 1880 by the Swedish engineer Jonas Wenström. Faraday's 1831 discovery of the principle of the AC transformer was not put to practical use until the late 1880s when the heated debate over the merits of direct-current and alternating-current systems for power transmission was settled in favour of the latter.

At first, the only serious consideration for electric power was arc lighting, in which a brilliant light is emitted by an electric spark between two electrodes. The arc lamp was too powerful for domestic use, however, and so it was limited to large installations like lighthouses, train stations, and department stores.

Commercial development of an incandescent filament lamp, first invented in the 1840s, was delayed until a filament could be made that would heat to incandescence without melting and until a satisfactory vacuum tube could be built. The mercury pump, invented in 1865, provided an adequate vacuum, and a satisfactory carbon filament was developed independently by the English physicist Sir Joseph Wilson Swan and the American inventor Thomas A. Edison during the late 1870s. By 1880 both had applied for patents for their incandescent

110

lamps, and the ensuing litigation between the two men was resolved by the formation of a joint company in 1883.

Thanks to the incandescent lamp, electric lighting became an accepted part of urban life by 1900. Since then, the tungsten filament lamp, introduced during the early 1900s, has become the principal form of electric lamp, though more efficient fluorescent gas discharge lamps have found widespread use as well.

Electricity took on a new importance with the development of the electric motor. This machine, which converts electric energy to mechanical energy, has become an integral component of a wide assortment of devices ranging from kitchen appliances and office equipment to industrial robots and rapid-transit vehicles. Although the principle of the electric motor was devised by Faraday in 1821, no commercially significant unit was produced until 1873. In fact, the first important AC motor, built by the Serbian-American inventor Nikola Tesla, was not demonstrated in the United States until 1888.

Tesla began producing his motors in association with the Westinghouse Electric Company a few years after DC motors had been installed in trains in Germany and Ireland. By the end of the 19th century, the electric motor had taken a recognizably modern form. Subsequent improvements have rarely involved radically new ideas; however, the introduction of better designs and new bearing, armature, magnetic, and contact materials has resulted in the manufacture of smaller, cheaper, and more efficient and reliable motors.

The modern communications industry is among the most spectacular products of electricity. Telegraph systems using wires and simple electrochemical or electromechanical receivers proliferated in western Europe and the United States during the 1840s. An operable cable was installed under the English Channel in 1865, and a pair of transatlantic cables was successfully laid a year later. By 1872 almost all of the major cities of the world were linked by telegraph.

Alexander Graham Bell patented the first practical telephone in the United States in 1876 and the first public telephone services were operating within a few years. In 1895 the British physicist Sir Ernest Rutherford advanced Hertz's scientific investigations of radio waves and transmitted radio signals for more than one kilometre. Guglielmo Marconi, an Italian physicist and inventor, established wireless communications across the Atlantic employing radio waves of approximately 300- to 3,000-metre wavelength in 1901. Broadcast radio transmissions were established during the 1920s.

111

Telephone transmissions by radio waves, the electric recording and reproduction of sound, and television were made possible by the development of the triode tube. This three-electrode tube, invented by the American engineer Lee De Forest, permitted for the first time the amplification of electric signals. Known as the Audion, this device played a pivotal role in the early development of the electronics industry.

The first telephone transmission via radio signals was made from Arlington, Va., to the Eiffel Tower in Paris in 1915; and a commercial radio telephone service between New York City and London was begun in 1927. Besides such efforts, most of the major developmental work of this period was tied to the radio and phonograph entertainment industries and the sound film industry.

Rapid progress was made toward transmitting moving pictures, especially in Great Britain; just before World War II, the British Broadcasting Corporation inaugurated the first public television service. Today, many regions of the electromagnetic spectrum are used for communications, including microwaves in the frequency range of approximately 7×10^9 hertz for satellite communication links and infrared light at a frequency of about 3×10^{14} hertz for optical fibre communications systems.

Until 1939 the electronics industry was almost exclusively concerned with communications and broadcast entertainment. Scientists and engineers in Britain, Germany, France, and the United States did initiate research on radar systems capable of aircraft detection and antiaircraft fire-control during the 1930s, however, and this marked the beginning of a new direction for electronics. During World War II and after, the electronics industry made strides paralleled only by those of the chemical industry. Television became commonplace; and a broad array of new devices and systems, most notably the electronic digital computer, emerged.

The electronic revolution of the last half of the 20th century has been made possible in large part by the invention of the transistor (1947) and such subsequent developments as the integrated circuit. This miniaturization and integration of circuit elements has led to a remarkable diminution in the size and cost of electronic equipment and an equally impressive increase in its reliability.

19.3 DEVELOPMENT OF JET ENGINES

Like many other inventions, jet engines were envisaged long before they became a reality. The earliest proposals were based on adaptations of piston engines and were usually heavy and complicated.

The first to incorporate a turbine design was conceived as early as 1921, and the essentials of the modern turbojet were contained in a patent in 1930 by Frank Whittle in England. His design was first tested in 1937 and achieved its first flight in May 1941. In Germany, parallel but completely independent work followed issuance of a patent in 1935. It proceeded more rapidly, and the very first flight of a turbojet-powered aircraft, a Heinkel HE-178, came in August 1939. By the end of World War II these prototype aircraft had developed into a few operational turbojet squadrons in the German, British, and U.S. air forces.

In the military area, jet fighter aircraft developed rapidly and were in use during the Korean War (1950-53), flying at speeds of 1,000 kilometres per hour. During the next decade they overcame the sound barrier and established normal operations up to more than twice the speed of sound (Mach 2). Bomber and transport jet aircraft were also able to reach and cruise at supersonic speeds.

The first civil jet transport, the British de Havilland Comet, flew in 1949, and regular transatlantic jet services were started in 1958 with the Comet 4 and the American Boeing 707. By 1974 more than 90 percent of hours flown throughout the world were flown by jets; the first supersonic airliner, the British-French Concorde, flying at more than twice the speed of sound, entered regular service in January 1976.

During the 1980s various major aircraft manufacturers undertook programs to develop fuel-saving propfan and unducted-fan propulsion systems. Some authorities believe that the next generation of commercial air transport may very well be powered by such advanced-technology propeller engines.

19.4 HIGH-PRESSURE PHENOMENA IN DIAMOND MAKING

While modest pressures (less than 1,000 atm) have long been used in the manufacture of plastics, in the synthesis of chromium dioxide for magnetic recording tape, and in the growth of large, high-quality quartz crystals, the principal application of high-pressure materials technology lies in the synthesis of diamond and other super-hard abrasives. Approximately 100 tons of synthetic diamonds are produced each year--a weight comparable to the total amount of diamond mined since biblical times.

For centuries diamonds had been identified only as an unusual mineral found in river gravels; scientists had no clear idea about their mode of origin until the late 1860s, when South African miners found diamond embedded in its native matrix, the high-pressure volcanic rock called kimberlite. Efforts to make diamond by subjecting graphitic carbon to high pressure began shortly after that historic discovery.

Prior to the work of Bridgman, sufficient laboratory pressures for driving the graphite-to-diamond transition had not been achieved. Bridgman's opposed-anvil device demonstrated that the necessary pressures could be sustained, but high temperatures were required to overcome the kinetic barrier to the transformation.

Following World War II, several industrial laboratories, including Allmanna Svenska Elektriska Aktiebolaget (ASEA) in Sweden and Norton Company and General Electric in the United States, undertook major efforts to develop a commercial process. Diamond was first synthesized in a reproducible, commercially viable experiment in December 1954, when Tracy Hall, working for General Electric, subjected a mixture of iron sulfide and carbon to approximately 6 GPa and 1,500° C in a belt-type apparatus. General Electric employees soon standardized the processes and discovered that a melted ferrous metal, which acts as a catalyst, is essential for diamond growth at these conditions.

19.5 ROADS AND HIGHWAYS SAFETY

Traffic police (or road patrols or highway police) help improve road safety and traffic flow by enforcing driving regulations. They also regulate traffic at the scene of an accident and investigate accidents. Traffic enforcement has been aided by the use of technology--cameras, radar, video, and inductance loops--to detect and record traffic offenders automatically.

An important aspect of traffic regulation and accident prevention is the control of excessive speed, which contributes significantly to the number and severity of road crashes. Speed is commonly measured by radar devices or by pacing with a patrol car. In crash investigations, the speed of the cars is determined by the length of skid marks. Another key factor in road accidents is the influence of alcohol and drugs.

114

Tests for intoxication are now widely conducted; the most common is the breath test, in which the driver blows into a device that analyzes the alcohol content of the breath and indicates the approximate blood alcohol level. Many authorities believe that 0.50 gram of alcohol per litre of blood is a realistic limit for ordinary motorists, but that zero levels should be demanded for critical operators such as drivers of public transport vehicles.

Road safety can also be built into the road. Divided roads are many times safer than two-way roads. Crash severity can be reduced by the use of "soft" signs and light poles and by guardrails and impact attenuators in front of fixed roadside objects such as bridge piers and the noses at the exit ramps of a freeway. Better road surfaces, alignments, signing, and marking improve driving conditions and increase road safety.

Nevertheless, about 90 percent of crashes are primarily due to human error. Many crashes have been attributed to simple inattention or failure to see warnings. Alcohol, fatigue, inexperience, aggression, and excessive risk taking are the most common crash causes involving behavioural changes in drivers. Lack of driving skills is rarely an issue; most drivers do not need training as much as they need education and experience. Meanwhile, road engineers must design road systems that attempt to reduce the frequency and impact of human error.

19.6 PAINT

Paint is a decorative and protective coating commonly applied to rigid surfaces as a liquid consisting of a pigment suspended in a vehicle, or binder. The vehicle, usually a resin dissolved in a solvent, dries to a tough film, binding the pigment to the surface.

Paint was used for pictorial and decorative purposes in the caves of France and Spain as early as 15,000 BC. The earliest pigments, which were natural ores such as iron oxide, were supplemented by 6000 BC in China by calcined (fired) mixtures of inorganic compounds and organic pigments; vehicles were prepared from gum Arabic, egg white, gelatine, and beeswax. By 1500 BC the Egyptians were using dyes such as indigo and madder to make blue and red pigments. The exploitation of linseed oil (a drying oil useful as a vehicle) and zinc oxide (a white pigment) in the 18th century brought a rapid expansion of the European paint industry.

115

The 20th century saw important developments in paint technology, including the introduction of synthetic polymers as vehicles and of synthetic pigments; a new understanding of the chemistry and physics of paints; and coating materials with greater fire retardancy, corrosion resistance, and heat stability. Most significant was a return to water-based paints in the form of latex paints that combine easy application and cleanup with reduced hazard of fire.

In modern paint manufacture, pigment particles are dispersed in the vehicle by cylindrical mills that tumble heavy metal or ceramic balls through the paint, or by sand grinders that circulate a suspension of sand through the paint at high speed.

The basic white pigments include zinc oxide, zinc sulfide, lithopone, and titanium dioxide. Most black pigments are composed of elemental carbon. Common red pigments include the minerals iron oxide, cadmium, and cuprous oxide and various synthetic organic pigments. Yellow and orange pigments include chromates, molybdates, and cadmium compounds. Blue and green pigments are either inorganic (synthetic ultramarines and iron blues) or organic (phthalocyanines). Extenders or fillers are sometimes added to paint to increase its spreadability and strength.

19.7 CONCRETE BUTTRESS AND MULTIPLE-ARCH DAMS

Unlike gravity dams, buttress dams do not rely entirely upon their own weight to resist the thrust of the water. Their upstream face, therefore, is not vertical but inclines about 25° to 45°, so the thrust of the water on the upstream face inclines toward the foundation. Embryonic buttresses existed in some Roman dams built in Spain, among them the Proserpina.

As technology advanced, dams with thin buttresses of reinforced concrete supporting inclined panels of similar construction were built. In today's buttress dams, less account is taken of effecting maximum economy in the use of concrete. The trend is to reduce the area of costly formwork necessary and to avoid use of steel reinforcement. With greater heights, modern buttress dams are inevitably less slender.

Several variations are possible in the design of the junction between the buttresses at the water face. Where no relative movement in the buttress foundations is anticipated, the design can link individual buttress heads rigidly, by means of arches, to form a multiple-arch dam. A recent Canadian example of this type is the 703-foot-high multiple-arch Daniel Johnson Dam on the Manicouagan River in

Quebec. The dam uses a total of 14 buttresses in its crest length of 4,297 feet; two very much larger buttresses support the structure over the original riverbed.

Where buttress foundations might yield, the design must allow some freedom of movement between the heads of the buttresses. This is normally achieved by enlarging the heads until they are almost in contact and then joining them with flexible seals. Thus joined, the heads present a solid face to the water. Such a design was used in the construction in the Farahnaz Pahlavi Dam in Iran. Built for the Tehran Regional Water Board, this dam has a maximum height of 351 feet and a crest length of nearly 1,181 feet.

A comparison between the Daniel Johnson multiple-arch dam and the Farahnaz Pahlavi buttress dam shows that the buttresses have to be placed much closer together than is necessary with a multiple-arch dam. This allows each buttress to be more slender, however, and spreads the load more easily over the foundation.

The detailed design at the bottom of the Farahnaz Pahlavi buttresses was necessitated by weak foundation conditions at the site and by the need to limit the length of each buttress to reduce its response to seismic action. By contrast, the Daniel Johnson buttresses could be founded individually, exploiting fully an important advantage of buttress dams over gravity dams--that of smaller uplift forces.

19.8 WATER TURBINE TECHNOLOGY

Experiments on the mechanics of reaction wheels conducted by the Swiss mathematician Leonhard Euler and his son Albert in the 1750s found application about 75 years later. In 1826 Jean-Victor Poncelet of France proposed the idea of an inward-flowing radial turbine, the direct precursor of the modern water turbine. This machine had a vertical spindle and a runner with curved blades that was fully enclosed. Water entered radially inward and discharged downward below the spindle.

A similar machine was patented in 1838 by Samuel B. Howd of the United States and built subsequently. Howd's design was improved on by James B. Francis, who added stationary guide vanes and shaped the blades so that water could enter shock-free at the correct angle. His runner design, which came to be known as the Francis turbine (see above), is still the most widely used for medium-high heads. Improved control was proposed by James Thomson, a Scottish engineer, who

117

added coupled and pivoted curved guide vanes to assure proper flow directions even at part load.

A radial outward-flow turbine had been proposed in 1824 by the French engineering professor Claude Burdin and his former student Benoît Fourneyron. This device had a vertical axis carrying a runner with curved blades through which the water left almost tangentially. Fixed guide vanes, curved in the opposite direction, were mounted in an annulus inside the runner. Unfortunately the design made it difficult to support the runner and to take power off the turbine wheel. The first successful version of the turbine was built by Fourneyron in 1827.

More than 100 such machines were subsequently built all over the world; they achieved efficiencies up to 75 percent at full load with heads up to 107 metres. In 1844 Uriah A. Boyden added an outlet diffuser to recover part of the kinetic energy exiting the device and thereby further improved efficiency. Outward-flow turbines, however, are inherently unstable, and speed control is difficult. Moreover, the construction of outward-flow turbines is very complex as compared to that of Francis-type runners, and this fact led to their eventually being supplanted by the latter.

Francis turbines were augmented by the development of the Pelton wheel (1889) for small flow rates and high heads and by propeller turbines, first built by Kaplan in 1913, for large flows at low heads. Kaplan's variable-pitch propeller turbine, which still bears his name, was manufactured after 1920. These units, together with the Deriaz mixed-flow turbine (invented in 1956), constitute the arsenal of modern water turbines.

By the mid-19th century, water turbines were widely used to drive sawmills and textile mill equipment, often through a complex system of gears, shafts, and pulleys. After the widespread adoption of the steam engine they did not, however, become a major factor in power generation until the advent of the electric generator made hydroelectric power possible.

The world's first hydroelectric central station was built in 1882 in Appleton, Wis., only three years after Thomas Edison's invention of the light bulb. Its output of 12.5 kilowatts was used to light two paper mills and a house. Thereafter hydroelectric power development spread rapidly, though even by 1910 most units delivered only a few hundred to a few thousand kilowatts. Installations with more than 100,000-kilowatt capacity were not built until the 1930s. One of the first large U.S. plants was installed at Hoover Dam on the Colorado River between Nevada and Arizona. It began operating in 1936 and eventually included 17 Francis turbines capable of delivering from 40,000 to 130,000 kilowatts of power, along with two 3,000-kilowatt Pelton wheels.

The first pumped storage plant with a capacity of 1,500 kilowatts was built near Schaffhausen, Switz., in 1909. It made use of a separate pump and turbine, resulting in a relatively large and only barely economical system. The first U.S. plant, built on the Rocky River in Connecticut in 1929, was also only marginally economical. In the United States major work on pumped-storage hydropower began in the mid-1950s, following the success of a plant at Flatiron, Colo. Built in 1954, this facility was equipped with a reversible-pump turbine having a capacity of 9,000 kilowatts.

In highly industrialized countries, such as the United States and the nations of Western Europe, most potential sites for hydropower have already been tapped. Environmental concerns relating to the impact of large dams on the upstream watercourse and to the possible effect on aquatic life add to the likelihood that only a few large hydraulic plants will be built in the future.

From about the 1940s to the early 1970s, many small U.S. hydroelectric facilities (primarily those of less than 1,000-kilowatt capacity) were, in fact, closed down because high maintenance and supervision costs made them uneconomical compared to power plants that burn fossil fuels. Even though the increase in fossil-fuel costs since 1973 has led to the rehabilitation of some of these abandoned plants, only a marked increase in fuel prices, coupled with specific needs for irrigation or flood control, is likely to lead to significant new hydroelectric plant construction.

It is estimated that about 75 percent of the potential waterpower in the contiguous United States has already been developed, with the drainage area of the Columbia River in the Pacific Northwest leading in both developed and potential additional power. As of the late 1980s, hydroelectric power met about 13 percent of the total demand for electrical energy in the United States, though this amounts to only 3 percent of the combined U.S. energy usage for mechanical power, heat, light, and refrigeration.

The above considerations do not necessarily apply to such remote areas as Alaska, northern Canada, and Siberia in Russia, or to developing nations in regions of the Himalayas, Africa, and South America. In these areas it is estimated that only 23 percent of the potential waterpower has been developed. For example, less than 1

119

percent of the estimated 167 million kilowatts available in Alaska has been harnessed to date. Other river basins with large remaining potential capacities include the Fraser River in Canada, the Orinoco in Venezuela, the Brahmaputra in India, and the Yenisey-Angara in Russia. Turbine capacities for some of these remote areas may possibly exceed the current maximum of 740,000 kilowatts per unit.

19.9 DEVELOPMENT OF SECURITY SYSTEMS

The origins of security systems are obscure, but techniques for protecting the household, such as the use of locks and barred windows, are very ancient. As civilizations developed, the distinction between passive and active security was recognized, and responsibility for active security measures was vested in police and fire-fighting agencies.

By the mid-19th century, private organizations such as those of Philip Sorensen in Sweden and Allan Pinkerton in the United States had also begun to build efficient large-scale security services. Pinkerton's organization offered intelligence, counterintelligence, internal security, investigative, and law enforcement services to private business and government. Until the advent of collective bargaining in the United States, strikebreaking was also a prime concern. The Sorensen organization, in contrast, moved toward a loss-control service for industry. It provided personnel trained to prevent and deal with losses from crime, fire, accident, and flood and established the pattern for security services in the United Kingdom and elsewhere in western Europe.

World Wars I and II brought an increased awareness of security systems as a means of protection against military espionage, sabotage, and subversion; such programs in effect became part of a country's national-security system. After World War II much of this apparatus was retained as a result of international tensions and defense-production programs and became part of an increasingly professionalized complex of security functions.

The development and diffusion of security systems and hardware in various parts of the world has been an uneven process. In relatively underdeveloped countries, or the underdeveloped parts of recently industrializing countries, security technology generally exists in rudimentary form, such as barred windows, locks, and elementary personnel security measures. In many such regions, however, facilities of large international corporations and sensitive government installations employ sophisticated equipment and techniques.

120

Since the 1960s, crime-related security systems have grown especially rapidly in most countries. Among contributing factors have been the increase in number of security-sensitive businesses; development of new security functions, such as protection of proprietary information; increasing computerization of sensitive information subject to unique vulnerabilities; improved reporting of crime and consequent wider awareness; and the need in many countries for security against violent demonstrations, bombings, and hijackings.

Security systems are becoming increasingly automated, particularly in sensing and communicating hazards and vulnerabilities. This situation is true in both crime-related applications, such as intrusion-detection devices, and fire-protection alarm and response (extinguishing) systems. Advances in miniaturization and electronics are reflected in security equipment that is smaller, more reliable, and more easily installed and maintained.

121

20. COMPUTER SCIENCE

20.1 ARTIFICIAL INTELLIGENCE

Artificial intelligence (AI) is an area of research that goes back to the very beginnings of computer science. The idea of building a machine that can perform tasks perceived as requiring human intelligence is an attractive one. The tasks that have been studied from this point of view include game playing, language translation, natural-language understanding, fault diagnosis, robotics, and supplying expert advice.

20.2 COMPUTER GRAPHICS

Computer graphics is the field that deals with display and control of images on the computer screen. Applications may be broken down into four major categories:

(1) design (computer-aided design [CAD] systems), in which the computer is used as a tool in designing objects ranging from automobiles to bridges to computer chips by providing an interactive drawing tool and an interface to simulation and analysis tools for the engineer;

(2) fine arts, in which artists use the computer screen as a medium to create images of impressive beauty, cinematographic special effects, animated cartoons, and television commercials;

(3) scientific visualization, in which simulations of scientific events--such as the birth of a star or the development of a tornado--are exhibited pictorially and in motion so as to provide far more insight into the phenomena than would tables of numbers; and

(4) human-computer interfaces.

Graphics-based computer interfaces, which enable users to communicate with the computer by such simple means as pointing to an icon with a handheld device known as a mouse, have allowed millions of ordinary people to control application programs like spreadsheets and word processors. Graphics technology also supports windows (display boxes) environments on the workstation or personal computer screen, which allow users to work with different applications simultaneously, one in each window.

Graphics also provide realistic interfacing to video games, flight simulators, and other simulations of reality or fantasy. The term

122

virtual reality has been coined to refer to interaction with a computer-simulated virtual world.

A challenge for computer science has been to develop algorithms for manipulating the myriad lines, triangles, and polygons that make up a computer image. In order for realistic on-screen images to be generated, the problems introduced in approximating objects as a set of planar units must be addressed.

Edges of objects are smoothed so that the underlying construction from polygons is not visible, and representations of surfaces are textured. In many applications, still pictures are inadequate, and rapid display of real-time images is required. Both extremely efficient algorithms and state-of-the-art hardware are needed to accomplish such real-time animation. Technical details of graphics displays are discussed in computer graphics.

20.3 SYSTEMS ENGINEERING, A DESIGN EXAMPLE

The design of the commercial transport plane is an example of a systems engineering problem. In such a design the aerodynamic lift, the drag of fuselage and wings, the control apparatus, the propulsion system, and such auxiliary hardware as the landing gear all interact substantially. One element cannot be disturbed without affecting the others; all elements and aspects of the total system, and the interactions among them, must be considered.

Thus, if designers make the fuselage fatter and the wings smaller in an effort to carry more payloads at the same or higher speeds, a new control system might be needed because of the changes produced in the overall mechanical and aerodynamic characteristics of the vehicle.

Stronger and heavier landing gear might be needed to withstand higher landing speeds. Almost surely, the new design would call for larger engines and fuel tanks to compensate for greater aerodynamic drag. Thus the designers would have lost ground in some respects and gained in others.

The new plane might be more useful for short flights when not much fuel must be carried but less useful for long ones. Obviously, the system objective--the kind of airplane actually wanted--must control the direction of any such study.

The study becomes more interesting if a possible advance in basic technology is considered, such as an improvement in propulsion or aerodynamics, and it is desired to determine how it might best be applied in a new airplane design.

123

The central systems engineering question then would probably encompass the relation between the available new plane characteristics and the needs of the existing air transportation system. Clearly, such an investigation can be made only by going to one of the upper levels in the systems hierarchy.

Finally, to operate the new airplane successfully, a whole series of supporting functions may be required, including routine checkout, maintenance, and spare parts supply, in addition to functions directly involved in the plane's flight.

Though, under normal circumstances, these might readily be handled by the existing operating staff, it is part of the user orientation of the systems approach that the systems engineer is expected to anticipate any new requirements and make sure they are properly planned for.

To make adequate comparisons between competing objectives, a logical frame of reference, broad enough to include both, is needed. Thus, the systems engineer may study many situations in the framework of more than one system or a whole hierarchy of systems of steadily increasing generality.

In the example of an airplane, the airplane itself is a possible system, as are the group of planes owned by one airline, the total number of airplanes in a particular country, and that nation's transportation facilities. Though the simplest system--the airplane itself--is a satisfactory reference for specific design problems, a more general framework may be needed to approach broader problems.

Thus the individual airplane designer may seek to ameliorate air-traffic congestion by improving airplane takeoff and landing characteristics, permitting better utilization of existing airports. The airlines in turn may suggest construction of more and better airports.

From the point of view of the transportation system as a whole, the best step might be to invest more money in high-speed rail facilities to carry part of the air-traffic load. In systems engineering the error of studying the problem within too narrow a framework is called the error of sub-optimization.

20.4 INTEGRATED CURCUIT

The integrated circuit (IC), also called MICROCIRCUIT, is an assembly of electronic components, fabricated as a single unit, in which active semiconductor devices (transistors and diodes) and passive devices (capacitors and resistors) and their interconnections are built up on a chip of material called substrate. The circuit thus consists of a

124

unitary structure with no connecting wires. The individual circuit elements are microscopic in size.

All elements of integrated circuits are fabricated in situ by an iterative process of lithographic definition, deposition, and etching on the common substrate (in most cases, silicon) in such a manner that the resulting interconnected elements perform the desired electrical circuit function.

Many ICs, typically about 1.5 square centimetres each, are fabricated simultaneously on silicon wafers up to 25 centimetres in diameter and subsequently sawed into individual chips (dies) prior to packaging. An IC thus produced is usually sealed in a plastic package with electrical leads that are internally connected by fine wires to output pads on the silicon die and that permit the packaged IC to be attached to a circuit card.

The integration of a large number of semiconductor devices on a single die of silicon is made possible by the high operational efficiency of the individual devices. Because of this, power dissipation is minimal and so too are the requirements for heat removal. The result is an IC of high dependability, which is further enhanced by the process of integration itself because the method of manufacturing the electrical interconnections between the devices and circuit elements by metal deposition and etching yields an extremely dependable circuit structure.

The number of devices integrated on a single tiny chip has increased from an initial few to nearly 100,000,000 as circuit elements with ever-smaller features are employed. This has, in turn, led to a progressive increase in complexity, as exemplified by the metal-oxide-semiconductor (MOS), the dynamic random-access memory (DRAM), and the MOS microprocessor.

Since manufacturing cost is proportional to area, the use of smaller features has not only led to an increase in complexity and resulting functionality but also has produced a decrease in cost per transistor and an increase in circuit performance. There are two reasons for the latter: (1) the performance of the individual active devices improves as their internal dimensions become smaller; and (2) the quality of the performance of the entire circuit improves as the active devices are positioned closer to one another. Large-volume applications for ICs have further lowered the manufacturing costs per IC. The cumulative result is that the integrated circuit has become the most pervasive technology of the 20th century.

It has provided the cornerstone of modern microelectronics and has promoted the development of the so-called information society. Applications of ICs range from their use in supercomputers--which are bringing about revolutionary advances in medical diagnosis, biotechnology, aeronautical and space engineering, telecommunications, and defence systems--to the development of new consumer products capable of bringing services and information to the home and office environment that otherwise would not have been possible.

20.5 ELECTRONIC EAVESDROPPING

This is the computerised aid in the act of electronically intercepting conversations without the knowledge or consent of at least one of the participants. Historically, the most common form of electronic eavesdropping has been wiretapping, which monitors telephonic and telegraphic communication. It is legally prohibited in virtually all jurisdictions for commercial or private purposes.

Great controversy has evolved over the use of this technique to detect crime or to gather evidence for criminal prosecution. Opponents assert that the legitimate governmental interest in curtailing crime does not outweigh the great potential for infringing upon constitutional or fundamental guarantees of citizenship, such as individual privacy and freedom from unreasonable searches and seizures.

Wiretapping activities date back to the beginnings of telegraphic communication. In the United States, state statutes forbidding the interception of messages were enacted as early as 1862. The tapping of telephone lines began in the 1890s and was approved for use by police officials in the Supreme Court case of *Olmstead* v. *United States* (1928).

Federal investigative authorities continue to engage in wiretapping, although in 1934 Congress enacted restraints that severely limited the use of intercepted material as admissible evidence in judicial proceedings. In the 1960s and '70s the Supreme Court sought to protect individuals from "unreasonable searches and seizures" by circumscribing prosecution based on electronic surveillance. Some U.S. states prohibit wiretapping completely, whereas others authorize its use pursuant to a valid court order. With the adoption of the Crime Control Act of 1968, Congress authorized the use of electronic surveillance for a variety of serious crimes, subject to strict judicial control.

In England permission to employ a wiretap is granted only in cases of serious offence when interceptions are likely to result in conviction and other methods of investigation have failed. In most other jurisdictions wiretapping is authorized under prescribed circumstances at the request of judicial, prosecuting, or police officials. A court order is ordinarily required, but in some countries, such as Denmark and Sweden, exceptions are recognized in urgent cases.

The typically vague standards governing the use of wiretapping have also provoked controversy with regard to other listening devices. Transistors, microcircuits, and lasers, all products of space-age technology, have revolutionized the art of electronic eavesdropping.

One group of the new investigative tools takes the shape of a ray gun that transmits radio waves or laser beams. The ray is directed at the object of the investigation from hundreds of feet away and can imperceptibly pick up a conversation and return it to the listener. The power necessary to transmit a laser beam for carrying voices many miles is extremely small, and a laser beam is more difficult to detect than radio signals.

The most efficient and least expensive form of listening device is a radio transmitter made out of integrated microcircuits. One hundred typical microcircuits can be made on a piece of material smaller and thinner than a postage stamp. A transmitter so constructed can be concealed in a playing card or behind wallpaper.

127

21. HISTORY OF ENERGY-CONVERSION TECHNOLOGY

21.1 ATTEMPTS TO HARNESS NATURAL FORMS OF ENERGY

Early humans first made controlled use of an external, non-animal energy source when they discovered how to use fire. Burning dried plant matter (primarily wood) and animal waste, they employed the energy from this biomass for heating and cooking. The generation of mechanical energy to supplant human or animal power came very much later--only about 2,000 years ago--with the development of simple devices to harness the energy of flowing water and of wind.

21.2 WATERWHEELS

The earliest machines were waterwheels, first used for grinding grain. They were subsequently adopted to drive sawmills and pumps, to provide the bellows action for furnaces and forges, to drive tilt hammers or trip-hammers for forging iron, and to provide direct mechanical power for textile mills. Until the development of steam power during the Industrial Revolution at the end of the 18th century, waterwheels were the primary means of mechanical power production, rivalled only occasionally by windmills. Thus, many industrial towns, especially in early America, sprang up at locations where water flow could be assured all year.

The oldest reference to a water mill dates to about 85 BC, appearing in a poem by an early Greek writer celebrating the liberation from toil of the young women who operated the querns (primitive hand mills) for grinding corn. According to the Greek geographer Strabo, King Mithradates VI of Pontus in Asia used a hydraulic machine, presumably a water mill, by about 65 BC.

Early vertical-shaft water mills drove querns where the wheel, containing radial vanes or paddles and rotating in a horizontal plane, could be lowered into the stream. The vertical shaft was connected through a hole in the stationary grindstone to the upper, or rotating, stone.

The device spread rapidly from Greece to other parts of the world, because it was easy to build and maintain and could operate in any fast-flowing stream. It was known in China by the 1st century AD, was used throughout Europe by the end of the 3rd century, and had reached Japan by the year 610. Users learned early that performance could be improved with a millrace and a chute that would direct the water to one side of the wheel.

A horizontal-shaft water mill was first described by the Roman architect and engineer Vitruvius about 27 BC. It consisted of an undershot waterwheel in which water enters below the centre of the wheel and is guided by a millrace and chute.

The waterwheel was coupled with a right-angle gear drive to a vertical-shaft grinding wheel. This type of mill became popular throughout the Roman Empire, notably in Gaul, after the advent of Christianity led to the freeing of slaves and the resultant need for an alternative source of power.

Early large waterwheels, which measured about 1.8 metres (six feet) in diameter, are estimated to have produced about three horsepower, the largest amount of power produced by any machine of the time. The Roman mills were adopted throughout much of medieval Europe, and waterwheels of increasing size, made almost entirely of wood, were built until the 13th century.

In addition to flowing stream water, ocean tides were used to drive waterwheels. Tidal water was allowed to flow into large millponds, controlled initially through lock-type gates and later through flap valves. Once the tide ebbed, water was let out through sluice gates and directed onto the wheel. Sometimes the tidal flow was assisted by building a dam across the estuary of a small river.

Although limited in operation to ebbing tide conditions, tidal mills were widely used by the 12th century. The earliest recorded reference to tidal mills is found in the *Domesday Book* (1086), which also records more than 5,000 water mills in England south of the Severn and Trent rivers. (Tidal mills also were built along the Atlantic coast in Europe and centuries later on the eastern seaboard of the United States and in Guyana, where they powered sugarcane-crushing mills.)

The first analysis of the performance of waterwheels was published in 1759 by John Smeaton, an English engineer. Smeaton built a test apparatus with a small wheel (its diameter was only 0.61 metre) to measure the effects of water velocity, as well as head and wheel speed. He found that the maximum efficiency (work produced divided by potential energy in the water) he could obtain was 22 percent for an undershot wheel and 63 percent for an overshot wheel (*i.e.,* one in which water enters the wheel above its centre).

In 1776 Smeaton became the first to use a cast-iron wheel, and two years later he introduced cast-iron gearing, thereby bringing to an end the all-wood construction that had prevailed since Roman times. Based on his model tests, Smeaton built an undershot wheel for the London Bridge waterworks that measured 4.6 metres wide and that had a diameter of 9.75 metres. The results of Smeaton's experimental work came to be widely used throughout Europe for designing new wheels.

During the mid-1700s a reaction waterwheel for generating small amounts of power became popular in the rural areas of England. In this type of device, commonly known as a Barker's mill, water flowed into a rotating vertical tube before being discharged through nozzles at the end of two horizontal arms. These directed the water out tangentially, much in the way that a modern rotary lawn sprinkler does. A rope or belt wound around the vertical tube provided the power takeoff.

Early in the 19th century Jean-Victor Poncelet, a French mathematician and engineer, designed curved paddles for undershot wheels to allow the water to enter smoothly. His design was based on the idea that water would run up the surface of the curved vanes, come to rest at the inner diameter, and then fall away with practically no velocity.

This design increased the efficiency of undershot wheels to 65 percent. At about the same time, William Fairbairn, a Scottish engineer, showed that breast wheels (*i.e.,* those in which water enters at the 10- or two-o'clock position) were more efficient than overshot wheels and less vulnerable to flood damage. He used curved buckets and provided a close-fitting masonry wall to keep the water from flowing out sideways. In 1828 Fairbairn introduced ventilated buckets in which gaps at the bottom of each bucket allowed trapped air to escape. Other improvements included a governor to control the sluice gates and spur gearing for the power takeoff.

During the course of the 19th century, waterwheels were slowly supplanted by water turbines. Water turbines were more efficient; design improvements eventually made it possible to regulate the speed of the turbines and to run them fast enough to drive electric generators. This fact notwithstanding, waterwheels gave way slowly, and it was not until the early 20th century that they became largely obsolescent. Yet, even today some waterwheels still survive; in the early 1970s there were more than 1,000 grain mills in use in Portugal alone. Equipped with submerged bearings, these modern waterwheels certainly are

21.3 COMPOSITE MATERIALS

The use of composite materials, similarly assisted in both design and application by the use of computers, has grown from the occasional application for a non-structural part (*e.g.,* a baggage compartment door) to the construction of complete airframes. These materials have the additional advantage in military technology of having a low observable (stealth) quality to radar.

Some aircraft of composite materials began to appear in the late 1930s and '40s; normally these were plastic-impregnated wood materials, the most famous (and largest) example of which is the Duramold construction of the eight-engine Hughes flying boat. A few production aircraft also used the Duramold construction materials and methods.

During the late 1940s, interest developed in fibreglass materials, essentially fabrics made up of glass fibres. By the 1960s, enough materials and techniques had been developed to make more extensive use possible. The term "composite" for this method of construction indicates the use of different materials that provide strengths, light weight, or other functional benefits when used in combination that they cannot provide when used separately. They usually consist of a fibre-reinforced resin matrix. The resin can be a vinyl ester, epoxy, or polyester, while the reinforcement might be any one of a variety of fibres, ranging from glass through carbon, boron, and a number of proprietary types.

To these basic elements, strength is sometimes added by the addition of a core material, making in effect a structural sandwich. A core can be made up of a number of plastic foams (polystyrene, polyurethane, or others), wood, honeycombs (multi-cellular structures) of paper, plastic, fabric or metal and other materials.

The desired final shape, in terms of both external appearance and the internal structure required for adequate strength, of a component made of composite materials can be arrived at by a variety of means. The simplest is the laying up of fibreglass sheets, much as is done in building a canoe, impregnating the sheets with a resin, and letting the resin cure. More sophisticated techniques involve fashioning the material into specific shapes by elaborate machinery. Some techniques require the use of male or female moulds or both, while others employ vacuum bags that allow the pressure of the atmosphere to press the parts into the desired shape.

The use of composite materials opened up whole new methods of construction and enabled engineers to create less expensive, lighter, and stronger parts of more streamlined shapes than had previously been feasible with wood or metal. Like the computer, the use of

Andreas Sofroniou

composites has spread rapidly throughout the industry and will be developed even further in the future.

The coincident arrival of the new technology in computers and composite materials influenced commercial air transportation, where aircraft larger than the Boeing 747 and faster than the Concorde are not only possible but inevitable. In the field of business aircraft, the new technologies have resulted in a host of executive aircraft with the most modern characteristics. These include the uniquely configured Beech Starship, which is made almost entirely of composite materials, and the Piaggio Avanti, which also has a radical configuration and employs primarily metal construction but includes a significant amount of composite material.

Commercial air transports are using composite materials in increasing amounts and may ultimately follow the pattern of the military services, where large aircraft like the Northrop B-2 are made almost entirely of advanced composite materials.

The previously mentioned considerations, combined with the advances in computers and composites, have completely revised the role of the homebuilt aircraft. While the homebuilt aircraft has always been a part of the aviation scene (the Wright Flyer was in fact a "homebuilt"), the designs were for years typically quite conventional, often using components from existing aircraft. Since the emergence of the Experimental Aircraft Association (founded 1953) in the United States, the homebuilt movement has operated in advance of the aviation industry, pioneering the use of computers and composites and, especially, radical configurations.

While there are many practitioners in the field, one man, the American designer Burt Rutan, epitomized this transition of the homebuilt movement from backyard to leading-edge status. Rutan, of Mojave, Calif., had a long series of successful designs, which reached the highest degree of recognition with the *Voyager* aircraft, in which his brother Dick Rutan and Jeana Yeager made a memorable non-stop, non-refuelled flight around the world in 1986.

Three other areas of civil aviation have benefited enormously from these advances in technology. The first of these are vertical-takeoff-and-landing aircraft, including helicopters. The second are sailplanes, which have reached new levels in structural and aerodynamic refinement. The third are the wide variety of hang gliders and ultra-light aircraft, as well as the smaller but more sophisticated aircraft that depend on human or solar power. Each of these has been vastly improved by contemporary advances in design and construction, and each holds great promise for the future.

21.4 HYDRAULICS

Hydraulics is the branch of science concerned with the practical applications of fluids, primarily liquids, in motion. It is related to fluid mechanics, which in large part provides its theoretical foundation. Hydraulics deals with such matters as the flow of liquids in pipes, rivers, and channels and their confinement by dams and tanks. Some of its principles apply also to gases, usually in cases in which variations in density are relatively small. Consequently, the scope of hydraulics extends to such mechanical devices as fans and gas turbines and to pneumatic control systems.

Liquids in motion or under pressure did useful work for man for many centuries before French scientist-philosopher Blaise Pascal and Swiss physicist Daniel Bernoulli formulated the laws on which modern hydraulic-power technology is based. Pascal's law, formulated in about 1650, states that pressure in a liquid is transmitted equally in all directions; *i.e*, when water is made to fill a closed container, the application of pressure at any point will be transmitted to all sides of the container. In the hydraulic press, Pascal's law is used to gain an increase in force; a small force applied to a small piston in a small cylinder is transmitted through a tube to a large cylinder, where it presses equally against all sides of the cylinder, including the large piston.

Bernoulli's law, formulated about a century later, states that energy in a fluid is due to elevation, motion, and pressure. and if there are no losses due to friction and no work done, the sum of the energies remains constant. Thus. velocity energy, deriving from motion, can be partly converted to pressure energy by enlarging the cross section of a pipe, which slows down the flow but increases the area against which the fluid is pressing.

Until the 19th century it was not possible to develop velocities and pressures much greater than those provided by nature, but the invention of pumps brought a vast potential for application of the discoveries of Pascal and Bernoulli. In 1882 the city of London built a hydraulic system that delivered pressurized water through street mains to drive machinery in factories. In 1906 an important advance in hydraulic techniques was made when an oil hydraulic system was installed to raise and control the guns of the Uss "Virginia." In the 1920s, self-contained hydraulic units consisting of a pump, controls,

133

and motor were developed, opening the way to applications in machine tools, automobiles, farm and earth-moving machinery, locomotives, ships, airplanes, and spacecraft.

In hydraulic-power systems there are five elements: the driver, the pump, the control valves, the motor, and the load. The driver may be an electric motor or an engine of any type. The pump acts mainly to increase pressure. The motor may be a counterpart of the pump, transforming hydraulic input into mechanical output. Motors may produce either rotary or reciprocating motion in the load.

The growth of fluid-power technology since World War II has been phenomenal. In the operation and control of machine tools, farm machinery, construction machinery, and mining machinery, fluid power can compete successfully with mechanical and electrical systems (*see* fluidics). Its chief advantages are flexibility and the ability to multiply forces efficiently; it also provides fast and accurate response to controls. Fluid power can provide a force of a few ounces or one of thousands of tons.

Hydraulic-power systems have become one of the major energy-transmission technologies utilized by all phases of industrial, agricultural, and defence activity. Modern aircraft, for example, use hydraulic systems to activate their controls and to operate landing gears and brakes. Virtually all missiles, as well as their ground-support equipment, utilize fluid power. Automobiles use hydraulic-power systems in their transmissions, brakes, and steering mechanisms. Mass production and its offspring, automation, in many industries have their foundations in the utilization of fluid-power systems.

22. UNUSUAL TECHNOLOGICAL METHODOLOGIES

22.1 ELECTRIC ORGAN

The ELECTROPHONIC ORGAN is a keyboard musical instrument in which tone is generated by electronic circuits and radiated by loudspeaker. This instrument, which emerged in the early 20th century, was designed as an economical and compact substitute for the much larger and more complex pipe organ.

The electronic organ resembles a spinet, or upright, piano in size and general shape. Most instruments of this general type rely upon electronic oscillators (circuits carrying an alternating current at a specific frequency) to produce their sound. Each oscillator is capable of frequency variation for different pitches and is capable of reproducing a single melodic line. The instrument's multiple oscillators make it capable of reproducing music having multiple parts, such as a fugue by Johann Sebastian Bach.

The 200-ton, keyboard-operated telharmonium, which used rotating electromagnetic tone-wheels to generate sound, was an important precursor to the electronic organ. Made in 1904 by the American inventor Thaddeus Cahill, it was exhibited in Massachusetts and New York in 1906 but lapsed into obscurity by World War I.

The first successful electronic organ was developed in 1928 in France by Edouard Coupleux and Armand Givelet. It used electronic oscillators in place of the pipes of a conventional organ and was operated with keyboards and a pedal board. Another notable early electronic organ was the Rangertone (1931), invented by Richard H. Ranger of the United States.

In 1934 the Orgatron was introduced by Frederick Albert Hoschke; in this organ, tone was generated by reeds that vibrated by electrically fan-blown air. with the vibrations picked up electrostatically and amplified.

One of the most important and well known of the electronic organs is the Hammond organ, a sophisticated instrument having two manuals, or keyboards, and a set of pedals operated by the feet. It was patented by its American inventor Laurens Hammond in 1934.

Unlike most other instruments of its type, it produces its sound through a complex set of rotary, motor-driven generators. By means of a series of controls affecting the harmonics, or component tones, of the sound, a great variety of timbres (tone colours) can be reproduced that

135

to some degree imitate the sound of other instruments, such as the violin, the flute, the oboe, and the orchestral percussion instruments.

By the 1960s organ manufacturers had expanded their technology, supplanting vacuum tubes with transistors and solid-state circuitry. Circuits and components designed to operate television and radio receivers and high-fidelity phonographs were adapted to produce music.

In the 1970s digital micro-circuitry was used to operate a computer organ. In this device, sounds are not created internally but have been pre-recorded (sampled) and stored in the computer from which they can later be retrieved.

Musical tones or shapes--recorded from conventional windblown pipe organs--are coded into digital form and may be re-created by a special computer at the touch of the keys and stops. Other devices have been used to control reverberation, pitch, and the attack or delay of a note.

22.2 MUSICAL INSTRUMENT

This refers to any device that produces a musical sound. The principal types of such instruments, classified by the method of producing sound, are percussion, stringed, keyboard, wind, and electronic.

Musical instruments are almost universal components of human culture: archaeology has revealed pipes and whistles in the Palaeolithic Period and clay drums and shell trumpets in the Neolithic age. It has been firmly established that the ancient city cultures of Mesopotamia, the Mediterranean, India, East Asia, and the Americas all possessed diverse and well-developed assortments of musical instruments, indicating that a long previous development must have existed.

As to the origin of musical instruments, however, there can be only conjecture. Some scholars have speculated that the first instruments were derived from such utilitarian objects as cooking pots (drums) and hunting bows (musical bows); others have argued that instruments of music might well have preceded pots and bows; while in the myths of cultures throughout the world the origin of music has frequently been attributed to the gods, especially in areas where music seems to have been regarded as an essential component of the ritual believed necessary for spiritual survival.

Whatever their origin, the further development of the enormously varied instruments of the world has been dependent on the interplay of four factors: available material, technological skills, mythic and symbolic preoccupations, and patterns of trade and migration. Thus, residents of Arctic regions use bone, skin, and stone to construct instruments; residents of the tropics have wood, bamboo, and reed available; while societies with access to metals and the requisite technology are able to utilize these malleable materials in a variety of ways.

Myth and symbolism play an equally important role. Herding societies, for example, which may depend on a particular species of animal not only economically but also spiritually, often develop instruments that look or sound like the animal or prefer instruments made of bone and hide rather than stone and wood, even when all the materials are available.

Finally, patterns of human trade and migration have for many centuries swept musicians and their instruments across seas and continents, resulting in constant flux, change, and cross-fertilization and adaptation.

The sound produced by an instrument can be affected by many factors, including the material from which the instrument is made, its size and shape, and the way that it is played. For example, a stringed instrument may be struck, plucked, or bowed, each method producing a distinctive sound.

A wooden instrument struck by a beater sounds markedly different from a metal instrument, even if the two instruments are otherwise identical. On the other hand, a flute made of metal does not produce a substantially different sound from one made of wood, for in this case the vibrations are in the column of air in the instrument.

The characteristic timbre of wind instruments depends on other factors, notably the length and shape of the tube. The length of the tube not only determines the pitch but also affects the timbre: the piccolo, being half the size of the flute, has a shriller sound. The shape of the tube determines the presence or absence of the "upper partials" (harmonic or non-harmonic overtones), which give colour to the single note.

This article discusses the evolution of musical instruments, their structure and methods of sound production, and the purposes for which they have been used. Although it focuses on the families of instruments that have been prominent in Western art music, it also includes coverage of non-Western and folk instruments.

Andreas Sofroniou

22.3 HISTORY OF BREWING

Before 6000 BC, beer was made from barley in Sumeria and Babylonia. Reliefs on Egyptian tombs dating from 2400 BC show that barley or partly germinated barley was crushed, mixed with water, and dried into cakes. When broken up and mixed with water, the cakes gave an extract that was fermented by micro-organisms accumulated on the surfaces of fermenting vessels.

The basic techniques of brewing came to Europe from the Middle East. The Roman historians Pliny (in the 1st century BC) and Tacitus (in the 1st century AD) reported that Saxons, Celts, and Nordic and Germanic tribes drank ale. In fact, many of the English terms used in brewing (malt, mash, wort, and ale) are Anglo-Saxon in origin. During the Middle Ages, the monastic orders preserved brewing as a craft. Hops were in use in Germany in the 11th century, and in the 15th century they were introduced into Britain from Holland.

In 1420, beer was made in Germany by a bottom fermentation process; before that, yeast rose to the top of the fermenting product and was allowed to overflow or was manually skimmed. Brewing was a winter occupation, and ice was used to keep beer cool during the summer months. Such beer came to be called lager (from German *lagern,* "to store"). The term lager is still used to denote beer produced from bottom-fermenting yeast, and the term ale is now used for top-fermented British types of beer.

The Industrial Revolution brought the mechanization of brewing. Better control over the process, with the use of the thermometer and saccharometer, was developed in Britain and transferred to the Continent, where the development of ice-making and refrigeration equipment in the late 19th century enabled lager beers to be brewed in summer.

In the 1860s the French chemist Louis Pasteur established many of the microbiological practices still used in brewing. The Danish botanist Emile Hansen devised methods for growing yeasts in culture free of other yeasts and bacteria. This pure-culture technology was quickly taken up by continental lager brewers but not until the 20th century by the ale brewers of Britain. Meanwhile, German-style bottom-fermented lagers fermented by pure yeast cultures became dominant in the Americas.

Brewing in the 20th century is a large-scale industry. Modern breweries use stainless-steel equipment and computer-controlled automated operations, and they package beer in metal casks, glass bottles, aluminum cans, and plastic containers. Beers are now exported worldwide and are produced under license in foreign countries.

22.4 OPTOELECTRONIC DEVICE SOLAR CELLS

The solar cell was the first optoelectronic device developed and was demonstrated by Daryl M. Chapin, Calvin S. Fuller, and Gerald L. Pearson in 1954 with a diffused silicon p-n junction. The solar cell is a large-area photodiode that "detects" the solar emission spectrum rather than a specific optical signal wavelength, as do photodiodes. The solar cell is unbiased, and the load is connected directly across the two terminals of the p-n junction. One of the most important parameters is the conversion efficiency, which is the ratio of the maximum power output to the incident power.

The first space satellites were electrically powered by silicon solar cells, and these cells continue to be an important long-duration power source for satellites. The solar cells originally used for this purpose were made with single-crystal silicon and had conversion efficiencies of about 15 percent. The efficiency of the single-crystal silicon solar cell is limited to about 20 percent, because the long-wavelength emission is not absorbed in the silicon or is not absorbed near enough to the p-n junction for the photo-generated carriers to be collected by the junction.

For ground applications, lower-cost solar cells have been developed by using large-grained, polycrystalline silicon with efficiencies near 16 percent instead of the more expensive single-crystal silicon. Even thin-film amorphous solar cells with efficiencies of about 12 percent have been investigated for further cost reduction. In each case, the lower cost is accompanied by reduced conversion efficiency. Other materials, such as aluminum gallium arsenide or gallium arsenide, have been used in applications where an increased conversion efficiency of more than 25 percent can justify the significantly greater cost.

The design of solar cells is influenced by the solar emission spectrum. The effect of the atmosphere on sunlight at the Earth's surface is defined by the air mass. The solar spectrum outside the atmosphere is the air mass zero (AM0). The solar spectrum at the Earth's surface for the minimum path length with the Sun directly overhead is AM1.

The sunlight is attenuated by the atmosphere owing to the absorption of infrared rays by ozone and scattering by clouds and airborne particles. For the Sun at an angle of 60° from the overhead position, the spectrum is AM2. The solar spectral irradiance is given by the power per unit wavelength.

In a high-efficiency single-crystal solar cell with an AM0 efficiency in the range of 20-21 percent, top contact is made via metal stripes that cover only 0.5 percent of the top surface. The rear aluminum contact layer acts as a highly effective reflector. The inverted pyramid structure along the top surface serves to reduce the front-surface reflection and combines with the rear reflector to form an efficent light-trapping scheme.

A more economical fabrication technology utilizing polycrystalline silicon involves cutting wafers from cast blocks. The silicon cell measures 10 centimetres square. These cells are connected in large solar panels for application on the Earth's surface to provide power at remote sites.

22.5 LOGISTICS IN THE NUCLEAR AGE

The dropping of the first atomic bombs in August 1945 seemed to inaugurate a new era in warfare, demanding radical changes in logistic systems and techniques. The bombs did, in truth, give birth to a new line of weaponry of unprecedented destructive power. Within a decade they were followed by the thermonuclear weapon, an even greater leap in destructive force.

Development of intercontinental ballistic missiles and nuclear-powered, missile-firing submarines a few years later extended the potential range of destruction to targets anywhere on the globe. The following decades saw dramatic developments in the offensive capabilities of nuclear weapons and also, for the first time, in defences against them. But the world moved into the late 20th century without any of the new nuclear weaponry having been used in anger.

Most warfare, moreover, was limited in scale and made little use of advanced technology. It produced only nine highly mobilized war economies: the two Koreas (1950-53), Israel (1956, 1967, 1973), North Vietnam (1965-75), Biafra (1967-70), Iran and Iraq (1980-88)--all except Israel pre-industrial Third World countries.

The first major conflict in this period, the war in Korea (1950-53), seemed in many ways an extension of the positional campaigns in World War II. It was fought largely with World War II weapons, in some cases improved versions, and with stocks of munitions left over from that conflict. United Nations forces had an excellent base in nearby Japan whose factories made a major contribution by rebuilding U.S. World War II material. UN air superiority kept both Japan and Pusan, South Korea's major port of entry, free from communist air attack.

UN forces thus were able to funnel through Pusan supply tonnages comparable to those handled by the largest ports in World War II and to concentrate depots and other installations in the Pusan area to a degree that would have been suicidal without air superiority. The communist supply system, although technically primitive, functioned well under UN air attack, moving troops and supplies by night, organizing local labour, and exploiting the Chinese soldier's famous ability to fight well under extreme privation.

By World War II standards, the Korean War was a limited conflict (except for the two Korean belligerents, on whose soil it was fought). It involved only a partial, or "creeping," economic mobilization in the United States and a modest mobilization of reserves. Yet this was no small war.

Over three years about 37.2 million measurement tons of cargo were poured into the South Korean ports, more than three-fourths of the amount shipped to U.S. Army forces in all the Pacific theatres in World War II. Combined UN forces reached a peak strength of almost one million men; communist forces were considerably larger.

22.6 SPACE EXPLORATION PREHISTORY TO SPUTNIK

The dream of flight into space is as old as astronomy. Once people learned that the lights in the night sky were actual bodies a long way off, they felt an urge to travel to them. Tales of fanciful flight to the Earth's nearest celestial neighbour, the Moon, may be traced back to the 2nd century AD, when the Greek rhetorician Lucian wrote a satirical account of such a fantastic journey.

But hundreds of years were to pass before the motions of the planets about the Sun and the immense distances to be travelled were appreciated. Awareness of the universe evolved slowly until tools of observation were developed. In the 17th century Galileo's use of the telescope to study the Moon added greatly to contemporary knowledge.

Within a few weeks he had mapped the major visible mountains and valleys of the Moon and concluded that it was a solid world. Turning his telescope to the planet Jupiter, he discovered four tiny specks that passed slowly across its face.

141

Galileo deduced that these moving dots were moons of the planet. He postulated further that their small apparent size in comparison to that of the Earth's Moon was explained by the great distance to Jupiter. Thus for the first time a scale of magnitude was established.

Further observation with the telescope confirmed the pattern of the solar system described by Copernicus a hundred years earlier. Johannes Kepler calculated the elliptical orbits of the planets. Later in the 17th century, Isaac Newton formulated his "laws of motion," which at last placed physics and astronomy on a firm theoretical foundation. Not until nearly a hundred years later were the first free balloon ascents made and the journey upward and away from Earth begun.

Three brilliant men, destined to be called rocket pioneers, were the first scientists to conceive pragmatically of space flight: the Russian Konstantin Eduardovich Tsiolkovsky, the American Robert Hutchings Goddard, and the German Hermann Oberth. Technology in the early 20th century, however, was a long way from the level required for rocket-powered flight.

Nonetheless, the theory and dynamics of such flights were rigorously studied. By the end of World War II, the German development of rocket propulsion for aircraft and guided missiles (notably the V-2) had reached a high level. With the German surrender in 1945, the United States as well as its Allies--Great Britain, France, and the Soviet Union--fell heir to the technical knowledge of rocket power developed by the Germans.

The technical director of the German missile effort, Wernher von Braun, and some 150 of his top aides surrendered to U.S. troops. Most emigrated to the United States, where they assembled and launched V-2 missiles that had been captured and shipped there. The Soviet Union carried out an unpublicized but extensive program that must have been very similar; Britain and France conducted smaller programs.

In both the United States and the Soviet Union the development of military missile technology was essential to the achievement of satellite flight. Preparations for the International Geophysical Year (IGY, 1957-58) stimulated discussion of the possibility of launching artificial Earth satellites for scientific investigations.

As a result, the planning committee for the IGY in 1954 passed a formal resolution calling attention to the desirability of using artificial satellites in the IGY program. Both the United States and the Soviet Union responded with announcements that they would prepare scientific satellites for launching during the IGY. While the United

States was still developing a satellite launch vehicle, the Soviet Union startled the world by placing Sputnik 1 in orbit.

The launch of Sputnik 1 was followed a month later by Sputnik 2 carrying a live dog named Laika, catching millions by surprise. The failure by the United States to launch its small (two kilograms) payload on December 6, 1957, heightened that nation's political discomfiture in view of its supposed advanced status in science.

Following debates on the necessity to achieve parity, the U.S. government established the National Aeronautics and Space Administration (NASA) in 1958. Since that time, NASA has conducted virtually all major aspects of the U.S. space program.

22.7 TITANIUM PROCESSING

The resistance of titanium to many corrosive environments, particularly oxidizing and chloride-containing process streams, has led to widespread industrial applications. Titanium is resistant to all natural environments, including natural waste products, body fluids, and salt and brackish water; to most salt solutions, including chlorides, bromides, iodides, and sulfides; and to most oxidizing acids, organic acids, and alkaline solutions.

When strength is not a major consideration, commercially pure titanium is the material of choice because of its lower cost, ease of fabrication, and good corrosion resistance. Alloys such as Ti-0.15Pd, Ti-0.3Mo-0.8Ni, and Ti-3A1-8V-6Cr-4Mo-4Zr can extend the usefulness of the metal to either higher temperatures or stronger concentrations of reducing acids and acidic salts. In recent years, more high-strength alloys have been utilized for corrosion applications.

For example, Ti-6Al-4V, a versatile alloy that was developed in the 1950s for the aerospace industry, has become a very important material for medical prostheses such as hip-joint replacements because of its strength-to-weight ratio and immunity to body fluids. Ti-3Al-8V-6Cr-4Mo-4Zr, an even stronger alloy, also has excellent resistance to high-temperature sour gas (natural gas containing hydrogen sulfide) and is therefore used in energy extraction for down-hole tubing and casings and for instrumentation.

Several industrial processes have been improved as a result of the availability of titanium. After titanium was introduced as a replacement for stainless-steel diffusion washers in the pulp and paper industry, the metal's excellent performance encouraged the design of

143

new displacement bleaching systems using up to 35 tons of titanium components.

Typical parts include diffusers, central shafts, scrapers, filtrate pumps, heat exchangers, packing boxes, and valves. In the early 1960s it was discovered that coating titanium with a platinum-group metal or metal oxide produced an anode (a negatively charged electrode) that was slow to corrode in electrolytic solutions. Coated titanium anodes soon replaced graphite anodes in the chlorine industry, resulting in lower costs and products of higher purity. Extensions of this technology are now being applied to electro-galvanizing and tin-coating processes.

Chemical-process industries utilize titanium heat exchangers to eliminate corrosion problems caused by cooling waters containing chloride and sulfide, and several benefits can accrue from employing titanium on the process side of heat exchangers as well. Because the metal is resistant to erosion corrosion, titanium vessels can be subjected to process liquids flowing at high rates, thereby eliminating the danger of biofouling. In addition, titanium is the only metal known to be completely resistant to all forms of biofouling corrosion.

These advantages, along with its light weight, make the metal desirable for heat exchangers in naval vessels and offshore oil platforms.

Titanium is gaining greater recognition in consumer applications, such as eyeglass frames, watches, sports equipment, jewellery, high-performance automobiles, and roofing. Other possible applications include valves for automobile engines, scrubbers for flue-gas desulfurization, marine and offshore risers, joints and fittings, and nuclear-waste storage and transportation casks.

22.8 BAKING

Baking is the process of cooking by dry heat, especially in some kind of oven. It is probably the oldest cooking method. Bakery products, which include bread, rolls, cookies, pies, pastries, and muffins, are usually prepared from flour or meal derived from some form of grain. Bread, already a common staple in prehistoric times, provides many nutrients in the human diet.

The earliest processing of cereal grains probably involved parching or dry roasting of collected grain seeds. Flavour, texture, and digestibility were later improved by cooking whole or broken grains with water, forming gruel or porridge. It was a short step to the baking of a layer of viscous gruel on a hot stone, producing primitive flat bread. More sophisticated versions of flat bread include the Mexican tortilla, made of processed corn, and the chapati of India, usually made of wheat.

Baking techniques improved with the development of an enclosed baking utensil and then of ovens, making possible thicker baked cakes or loaves. The phenomenon of fermentation, with the resultant lightening of the loaf structure and development of appealing flavours, was probably first observed when doughs or gruels, held for several hours before baking, exhibited spoilage caused by yeasts.

Some of the effects of the microbiologically induced changes were regarded as desirable, and a gradual acquisition of control over the process led to traditional methods for making leavened bread loaves. Early baked products were made of mixed seeds with a predominance of barley, but wheat flour, because of its superior response to fermentation, eventually became the preferred cereal among the various cultural groups sufficiently advanced in culinary techniques to make leavened bread.

Brewing and baking were closely connected in early civilizations. Fermentation of a thick gruel resulted in a dough suitable for baking; a thinner mash produced a kind of beer. Both techniques required knowledge of the "mysteries" of fermentation and a supply of grain. Increasing knowledge and experience taught the artisans in the baking and brewing trades that barley was best suited to brewing, while wheat was best for baking.

By 2600 BC the Egyptians, credited with the first intentional use of leavening, were making bread by methods similar in principle to those of today. They maintained stocks of sour dough, a crude culture of desirable fermentation organisms, and used portions of this material to inoculate fresh doughs. With doughs made by mixing flour, water, salt, and leaven, the Egyptian baking industry eventually developed more than 50 varieties of bread, varying the shape and using such flavouring materials as poppy-seed, sesame, and camphor. Samples found in tombs are flatter and coarser than modern bread.

The Egyptians developed the first ovens. The earliest known examples are cylindrical vessels made of baked Nile clay, tapered at the top to give a cone shape and divided inside by a horizontal shelf-like partition. The lower section is the firebox; the upper section is the baking chamber. The pieces of dough were placed in the baking chamber through a hole provided in the top.

In the first two or three centuries after the founding of Rome, baking remained a domestic skill with few changes in equipment or processing methods. According to Pliny the Elder, there were no bakers in Rome until the middle of the 2nd century BC. As well-to-do families increased, women wishing to avoid frequent and tedious bread making began to patronize professional bakers, usually freed slaves. Loaves moulded by hand into a spheroidal shape, generally weighing about a pound, were baked in a beehive-shaped oven fired by wood. *Panis artopticius* was a variety cooked on a spit, *panis testuatis* in an earthen vessel.

Although Roman professional bakers introduced technological improvements, many were of minor importance, and some were essentially reintroductions of earlier developments. The first mechanical dough mixer, attributed to Marcus Virgilius Euryasaces, a freed slave of Greek origin, consisted of a large stone basin in which wooden paddles, powered by a horse or donkey walking in circles, kneaded the dough mixture of flour, leaven, and water.

Guilds formed by the miller-bakers of Rome became institutionalized. During the 2nd century AD, under the Flavians, they were organized into a "college" with work rules and regulations prescribed by government officials. The trade eventually became obligatory and hereditary, and the baker became a kind of civil servant with limited freedom of action.

During the early Middle Ages, baking technology advances of preceding centuries disappeared, and bakers reverted to mechanical devices used by the ancient Egyptians and to more backward practices. But in the later Middle Ages the institution of guilds was revived and expanded. Several years of apprenticeship were necessary before an applicant was admitted to the guild; often an intermediate status as journeyman intervened between apprenticeship and full membership (master). The rise of the bakers' guilds reflected significant advances in technique.

A 13th-century French writer named 20 varieties of bread varying in shape, flavourings, preparation method, and quality of the meal used. Guild regulations strictly governed size and quality. But outside the cities bread was usually baked in the home.

In medieval England rye was the main ingredient of bread consumed by the poor; it was frequently diluted with meal made from other cereals or leguminous seeds. Not until about 1865 did the cost of white bread in England drop below brown bread.

At that time improvements in baking technology began to accelerate rapidly, owing to the higher level of technology generally. Ingredients of greater purity and improved functional qualities were developed,

146

along with equipment reducing the need for individual skill and eliminating hand manipulation of bread doughs.

Automation of mixing, transferring, shaping, fermentation, and baking processes began to replace batch processing with continuous operations. The enrichment of bread and other bakery foods with vitamins and minerals was a major accomplishment of the mid-20th-century baking industry.

22.9 TOYS

These refer to plaything for an infant or child. A toy is often an instrument used in a game. Toys, playthings, and games survive from the most remote past and from a great variety of cultures. They vary from the simplest to the most complex, from the natural stick selected by a child and imagined to be a hobbyhorse to sophisticated and complex mechanical devices that entertained both young and old in the courts of 18th-century Europe.

Museums in many countries exhibit antique objects, the original purpose of which may be unknown but which children may have used or adopted for playthings. A clay animal figure on wheels, from an early Mexican culture in which no other record of the wheel has been discovered, may have been a toy. One of the most ancient toys is the ball.

Play with toys follows two main directions, imitative and instructive. The earliest types of play probably developed from the instinct for self-preservation. In many human cultures, one of the first things taught to the young was the use of weapons, and the natural club or stick was the prototype of toy swords, guns, tanks, airplanes, ships, and other military instruments of play.

Most games and sports requiring physical action derived from practice of the skills of warfare, and the instruments of the game or sport were regarded as weapons. Toy soldiers and weapons dating from the Middle Ages are extant. The latest developments in warfare are represented among contemporary toys, as are those weapons and machines fantasized in science fiction books and motion pictures.

A basic toy is the doll. Every epoch and culture has provided its children with miniatures of human beings or animals and of the artefacts used in daily living. Many static toys are of this type: miniature versions of real beings or objects that lend themselves to an imaginative or imitative use.

147

Moving toys include a wider variety. It is probable that many experiments with basic physical principles were first realized in the form of moving toys known through literary description. Explosive toy weapons and rockets developed from the early use of gunpowder for fireworks by the Chinese. Balance and counterbalance, the wheel, the swing, the pendulum, flight, centrifugal force, magnetism, the spring, and a multitude of other devices and principles have been utilized in toys.

Modern technological developments have made possible the production of such sophisticated moving toys as scale-model electric railroad trains and auto-racing tracks and cars, radio-controlled model aircraft, and dolls that walk, talk, and perform other stunts when activated by a beam of light.

Coordination and other manual skills develop from cumulative childhood experiences received during the manipulation of toys-- marbles, jackstones, and other toys requiring use of hands and bodies. Mental agility, beginning with childhood, is challenged by puzzles of spatial relationships.

Roman children and adults threw knucklebones, which were probably the precursors of dice as well as of jackstones. Dice, in turn, are essential in a host of other games of chance. Other forms of toys probably derive from magical artefacts and fetishes that played a prominent part in primitive religions. In celebrating the Mexican festival of the Day of the Dead, sugar is formed into elaborate and beautiful skulls, tombs, and angels; they are essentially religious symbols, but in the hands of children they become toys that are played with and finally eaten.

Christmas-tree decorations, Easter eggs, the Neapolitan *presepio* (crèche) with its wealth of elaborate figures representing the birth of Jesus are other examples of toys of religious origin. A modern relic of early culture, the kachina doll of the Pueblo Indians, while essentially an instructive sacred object, inevitably is played with by children.

Under the pressure of industrialization, folk culture and tradition are rapidly disappearing, but in many countries a variety of folk or homemade toys can still be found. Toys in developed countries are usually mass-produced, with technology providing their locomotion and other actions.

22.10 WARNING SYSTEM

148

History abounds with examples of successful military surprises; examples of effective warning are difficult to find. Military training emphasized the value of surprise, stratagem, and deception, but the value of warning was long neglected. Flank and rear guards, to protect marching columns, patrols and scouts to locate the enemy, and sentries to guard camps, were of course used in war from earliest times.

Animals were sometimes employed to detect the approach of an enemy; dogs and horses were especially favoured, though, according to the ancient historian Livy, the Romans used geese to detect the night attack of the Gauls on Rome in the 4th century BC. High ground, favourable for observation, was often supplemented by watchtowers, such as those placed along the Great Wall of China and on Hadrian's Wall in Britain.

The observation balloon was an important technological advance. First used in warfare by the French in the late 18th century, primarily for offensive reconnaissance on the battlefield, its defensive possibilities were demonstrated in the American Civil War; in May 1863 a balloon of the army of the Potomac detected Lee's army moving from its camp across the Rappahannock to commence the Gettysburg campaign. Aerial photography had already been pioneered by the French and used in the War of Italian Independence (1859).

A balloon observer in the Spanish-American War of 1898 is credited with discovering an alternate route up San Juan Hill during the battle there. A few other successes are ascribed to such observation before the balloon was supplemented by the far more valuable airplane in World War I. Nevertheless, the balloon never fulfilled its potential as a warning device.

In sea warfare, warning and detection were equally neglected. As far back as the Minoan civilization of Crete, patrol ships were used, but mainly for offensive purposes. In later centuries, raised quarterdecks and lookout posts atop sailing masts were provided, but the beginnings of serious maritime detection technology did not come until the advent of the submarine.

Binoculars, telescopes, the telegraph, and telephone were well established military equipment by 1914; the airplane, first used by the Italians in the Italo-Turkish War of 1911, showed its potential as an observation device at the Battle of the Marne. Radio communications provided the means to make air observations immediately available. Aerial combat became inevitable as each side tried to deny the other its aerial reconnaissance.

149

Searchlights, first used in the Russo-Japanese War (1904), saw large-scale use in World War I to detect dirigibles and aircraft on night bombardment missions. Flares were used to illuminate the battlefield between trenches to detect raiding parties. Listening devices, using directional horns to detect and locate enemy aircraft, were also used with limited success.

Despite the novelties of World War I, World War II produced far more technological innovation. Radar made obsolete the slow and inaccurate older listening devices. Radio communications made great strides, particularly in the very high frequency range. The combination of radar and interference-free very high frequency communications was pivotal in permitting the RAF to resist Hitler's aerial attack and win the Battle of Britain.

Notwithstanding radar sophistication, ground spotters played an important role in filling the gaps between radar coverage. Their messages, forwarded to a plotting centre, were assembled to trace the progress of intruders (tracking).

The advent of nuclear weapons (1945), especially when coupled later with the speed and range of intercontinental missiles, gave new dimensions to the value of surprise for the attacker. Long-term warning was suddenly of paramount importance. Not only did all forms of unequivocal warning become indispensable but the warning had to be made credible to an aggressor; that is, an assurance had to be given that the retaliatory weapons would not all be destroyed by a first strike. Bomber aircraft were kept in the air to avoid destruction on the ground and attempts were made to provide a degree of protection for the civilian population through shelters.

Practically all aspects of science and technology have been introduced into today's warfare and warning systems: airplanes, helicopters, submarines, earth satellites, television, lasers, and magnetic, acoustic, seismic, infrared, nuclear, and chemical detectors.

23. FAMILIAR TECHNOLOGICAL DEVICES

23.1 ROMAN ROADS

The greatest systematic road builders of the ancient world were the Romans, who were very conscious of the military, economic, and administrative advantages of a good road system. The Romans drew their expertise mainly from the Etruscans--particularly in cement technology and street paving--though they probably also learned skills from the Greeks (masonry), Cretans, Carthaginians (pavement structure), Phoenicians, and Egyptians (surveying). Concrete made from cement was a major development that permitted many of Rome's construction advances.

The Romans began their road-making task in 334 BC and by the peak of the empire had built nearly 53,000 miles of road connecting their capital with the frontiers of their far-flung empire. Twenty-nine great military roads, the *viae militares*, radiated from Rome. The most famous of these was the Appian Way. Begun in 312 BC, this road eventually followed the Mediterranean coast south to Capua and then turned eastward to Beneventum, where it divided into two branches, both reaching Brundisium (Brindisi). From Brundisium the Appian Way traversed the Adriatic coast to Hydruntum, a total of 410 miles from Rome.

The typical Roman road was bold in conception and construction. Where possible, it was built in a straight line from one sighting point to the next, regardless of obstacles, and was carried over marshes, lakes, ravines, and mountains. In its highest stage of development, it was constructed by excavating parallel trenches about 40 feet apart to provide longitudinal drainage--a hallmark of Roman road engineering. The foundation was then raised about three feet above ground level, employing material taken from the drains and from the adjacent cleared ground.

As the importance of the road increased, this embankment was progressively covered with a light bedding of sand or mortar on which four main courses were constructed: (1) the *statumen* layer 10 to 24 inches (250 to 600 millimetres) thick, composed of stones at least 2 inches in size, (2) the *rudus*, a 9-inch-thick layer of concrete made from stones under 2 inches in size, (3) the *nucleus* layer, about 12 inches thick, using concrete made from small gravel and coarse sand, and, for very important roads, (4) the *summum dorsum*, a wearing surface of large stone slabs at least 6 inches deep.

The total thickness thus varied from 3 to 6 feet. The width of the Appian Way in its ultimate development was 35 feet. The two-way, heavily crowned central carriageway was 15 feet wide. On each side it was flanked by curbs 2 feet wide and 18 inches high and paralleled by one-way side lanes 7 feet wide. This massive Roman road section, adopted about 300 BC, set the standard of practice for the next 2,000 years.

The public transport of the Roman Empire was divided into two classes:

(1) *cursus rapidi*, the express service, and

(2) *agnarie*, the freight service.

In addition, there was an enormous amount of travel by private individuals. The most widely used vehicles were the two-wheeled chariot drawn by two or four horses and its companion, the cart used in rural areas.

A four-wheeled *raeda* in its passenger version corresponded to the stage coaches of a later period and in its cargo version to the freight wagons. Fast freight *raedae* were drawn by 8 horses in summer and 10 in winter and, by law, could not haul in excess of 750 pounds, or 330 kilograms. Speed of travel ranged from a low of about 15 miles per day for freight vehicles to 75 miles per day by speedy post drivers.

23.2 DRESS, NATURE AND PURPOSES

Perhaps the most obvious function of dress is to provide warmth and protection. Many scholars believe, however, that the first crude garments and ornaments worn by humans were designed not for utilitarian but for religious or ritual purposes. Other basic functions of dress include identifying the wearer (by providing information about sex, age, occupation, or other characteristic) and making the wearer appear more attractive.

Although it is clear why such uses of dress developed and remain significant, it can often be difficult to determine how they are achieved. Some garments thought of as beautiful offer no protection whatsoever and may in fact even injure the wearer. Items that definitely identify one wearer can lose their meaning in another time and place. Clothes that are deemed handsome in one period are declared downright ugly in the next, and even uniforms--the simplest and most easily identified costume--are subject to change.

What are the reasons for such changes? Why do people replace useful, attractive garments before they are worn out? In short, why does fashion, as opposed to mere dress, exist?

There are no simple answers to such questions, of course, and any one reason is influenced by a multitude of others, but certainly one of the most prevalent theories is that fashion serves as a reflection of social and economic standing. Thus, in relatively static societies with limited movement between classes, as in many parts of Asia until modern times or in Europe before the Middle Ages (or later in some areas), styles generally did not undergo major or rapid change.

In contrast, when lower classes have the ability to copy upper classes, the upper classes quickly instigate fashion changes that demonstrate their authority and high position. During the 20th century, for example, improved communication and manufacturing technology enabled new styles to trickle down from the elite to the masses at ever faster speeds, with the result that more styles were introduced than at any other time.

Furthermore, the idea that fashion is a reflection of wealth and prestige can be used to explain the popularity of many styles throughout costume history. For example, courts have been a major source of fashion in the West, and clothes that are difficult to obtain and expensive to maintain have frequently been at the forefront of fashion. Ruffs, for example, required servants to reset them with hot irons and starch every day and so were not generally worn by ordinary folk.

As such garments become easier to buy and care for, they lose their exclusivity and hence much of their appeal. For the same reason, when fabrics or materials are rare or costly, styles that require them in excessive, extravagant amounts become particularly fashionable--as can be seen in the 16th-century vogue for slashing outer garments to reveal a second layer of luxurious fabric underneath.

Similarly, impractical fashions that clearly demonstrate the wearer does not need to work, and indeed would find it difficult to do so dressed in such a manner, have often been considered beautiful. Examples include the Chinese practice of binding aristocratic women's feet, making it impossible for the women to walk far, and the recurrent popularity in Europe of styles that limited a woman's ability to manoeuvre or move by confining her into frequently injurious corsets and weighting her down with excessive layers of petticoats and skirts.

Women have traditionally been the targets of the most extreme forms of impractical fashion because they have frequently been viewed as little more than a frivolous ornament for a man's arm or household. The fact that a woman is dressed in such a manner proves not only that she does not work but also that her husband or father can afford to hire servants to work for her.

Men have worn their share of impractical clothing, however. The late Gothic houppelande, for example, a courtly style worn by both sexes, was far too voluminous for peasants to work in, even if they could have afforded all the material necessary for its manufacture. The best illustrations of the new garment are found in *Les Très Riches Heures du duc de Berry*, at the Condé Museum, Chantilly, Fr. These show that the duke wore the houppelande down to the floor, but his servitors, who needed to move more freely, wore shorter gowns. Length thus provided an immediate signal of status.

The foregoing discussion does not attempt to be a comprehensive introduction to even one influence on fashion; it merely tries to suggest some of the ways in which costume can be analyzed and interpreted. Similar treatments of four other factors affecting fashion follow.

23.3 MICROWAVES

The microwave region extends from 1,000 to 300,000 MHz (or 30-centimetre to one-millimetre wavelengths). Although microwaves were first produced and studied in 1886 by Hertz, their practical application had to await the invention of suitable generators, such as the klystron and magnetron.

Microwaves are the principal carriers of high-speed telegraphic data transmissions between stations on the Earth and also between ground-based stations and satellites and space probes. A system of synchronous satellites about 36,000 kilometres above the Earth is used for international broadband telegraphy of all kinds of communications--*e.g.,* television, telephone, and tele-facsimile (FAX).

Microwave transmitters and receivers are parabolic dish antennas. They produce microwave beams whose spreading angle is proportional to the ratio of the wavelength of the constituent waves to the diameter of the dish. The beams can thus be directed like a searchlight. Radar beams consist of short pulses of microwaves. One can determine the distance of an airplane or ship by measuring the time it takes such a pulse to travel to the object and, after reflection, back to the radar dish antenna.

Moreover, by making use of the change in frequency of the reflected wave pulse caused by the Doppler effect (see above Speed of electromagnetic radiation and the Doppler effect), one can measure the speed of objects. Microwave radar is therefore widely used for guiding airplanes and vessels and for detecting speeding motorists. Microwaves can penetrate clouds of smoke, but are scattered by water droplets, and so are used for mapping meteorologic disturbances and in weather forecasting (see weather forecasting).

Microwaves play an increasingly wide role in heating and cooking food. They are absorbed by water and fat in foodstuffs (*e.g.*, in the tissue of meats) and produce heat from the inside. In most cases, this reduces the cooking time a hundredfold. Such dry objects as glass and ceramics, on the other hand, are not heated in the process, and metal foils are not penetrated at all.

The heating effect of microwaves destroys living tissue when the temperature of the tissue exceeds 43° C (109° F). Accordingly, exposure to intense microwaves in excess of 20 milliwatts of power per square centimetre of body surface is harmful. The lens of the human eye is particularly affected by waves with a frequency of 3,000 MHz, and repeated and extended exposure can result in cataracts.

Radio waves and microwaves of far less power (microwatts per square centimetre) than the 10-20 milliwatts per square centimetre needed to produce heating in living tissue can have adverse effects on the electrochemical balance of the brain and the development of a fetus if these waves are modulated or pulsed at low frequencies between 5 and 100 hertz, which are of the same magnitude as brain wave frequencies.

Various types of microwave generators and amplifiers have been developed. Vacuum-tube devices, the klystron and the magnetron, continue to be used on a wide scale, especially for higher-power applications. Klystrons are primarily employed as amplifiers in radio relay systems and for dielectric heating, while magnetrons have been adopted for radar systems and microwave ovens.

Solid-state technology has yielded several devices capable of producing, amplifying, detecting, and controlling microwaves. Notable among these are the Gunn diode and the tunnel (or Esaki) diode. Another type of device, the maser (acronym for "microwave amplification by stimulated emission of radiation") has proved useful in such areas as radio astronomy, microwave radiometry, and long-distance communications.

Astronomers have discovered what appear to be natural masers in some interstellar clouds. Observations of radio radiation from interstellar hydrogen (H_2) and certain other molecules indicate amplification by the maser process. Also, as was mentioned above, microwave cosmic background radiation has been detected and is considered by many to be the remnant of the primeval fireball postulated by the big-bang cosmological model.

23.4 INSTRUMENTATION

Instrumentation in technology is the development and use of precise measuring equipment. Although the sensory organs of the human body can be extremely sensitive and responsive, modern science and technology rely on the development of much more precise measuring and analytical tools for studying, monitoring, or controlling all kinds of phenomena.

Some of the earliest instruments of measurement were used in astronomy and navigation. The armillary sphere, the oldest known astronomical instrument, consisted essentially of a skeletal celestial globe whose rings represent the great circles of the heavens. The armillary sphere was known in ancient China; the ancient Greeks were also familiar with it and modified it to produce the astrolabe, which could tell the time or length of day or night as well as measure solar and lunar altitudes.

The compass, the earliest instrument for direction finding that did not make reference to the stars, was a striking advance in instrumentation made about the 11th century. The telescope, the primary astronomical instrument, was invented about 1608 by the Dutch optician Hans Lippershey and first used extensively by Galileo.

Instrumentation involves both measurement and control functions. An early instrumental control system was the thermostatic furnace developed by the Dutch inventor Cornelius Drebbel (1572-1634), in which a thermometer controlled the temperature of a furnace by a system of rods and levers. Devices to measure and regulate steam pressure inside a boiler appeared at about the same time. In 1788 the Scotsman James Watt invented a centrifugal governor to maintain the speed of a steam engine at a predetermined rate.

Instrumentation developed at a rapid pace in the Industrial Revolution of the 18th and 19th centuries, particularly in the areas of dimensional measurement, electrical measurement, and physical analysis. Manufacturing processes of the time required instruments capable of achieving new standards of linear precision, met in part by the screw micrometer, special models of which could attain a precision of 0.000025 mm (0.000001 inch).

The industrial application of electricity required instruments to measure current. voltage, and resistance. Analytical methods, using such instruments as the microscope and the spectroscope, became increasingly important; the latter instrument, which analyzes by wave length the light radiation given off by incandescent substances, began to be used to identify the composition of chemical substances and stars.

In the 20th century the growth of modern industry, the introduction of computerization, and the advent of space exploration have spurred still greater development of instrumentation, particularly of electronic devices. Often a transducer, an instrument that changes energy from one form into another (such as the photocell, thermocouple, or microphone) is used to transform a sample of the energy to be measured into electrical impulses that are more easily processed and stored.

The introduction of the electronic computer in the 1950s, with its great capacity for information processing and storage, virtually revolutionized methods of instrumentation, for it allowed the simultaneous comparison and analysis of large amounts of information. At much the same time, feedback systems were perfected in which data from instruments monitoring stages of a process are instantaneously evaluated and used to adjust parameters affecting the process. Feedback systems are crucial to the operation of automated processes.

Most manufacturing processes rely on instrumentation for monitoring chemical, physical, and environmental properties, as well as the performance of production lines. Instruments to monitor chemical properties include the refractometer, infrared analyzers, chromatographs, and pH sensors.

A refractometer measures the bending of a beam of light as it passes from one material to another; such instruments are used, for instance, to determine the composition of sugar solutions or the concentration of tomato paste in catsup. Infrared analyzers can identify substances by the wavelength and amount of infrared radiation that they emit or reflect.

Chromatography, a sensitive and swift method of chemical analysis used on extremely tiny samples of a substance, relies on the different rates at which a material will adsorb different types of molecules. The acidity or alkalinity of a solution can be measured by pH sensors.

Instruments are also used to measure physical properties of a substance, such as its turbidity, or amount of particulate matter in a solution. Water purification and petroleum-refining processes are

157

monitored by a turbidimeter, which measures how much light of one particular wavelength is absorbed by a solution. The density of a liquid substance is determined by a hydrometer, which measures the buoyancy of an object of known volume immersed in the fluid to be measured.

The flow rate of a substance is measured by a turbine flowmeter, in which the revolutions of a freely spinning turbine immersed in a fluid are measured, while the viscosity of a fluid is measured by a number of techniques, including how much it dampens the oscillations of a steel blade.

Instruments used in medicine and biomedical research are just as varied as those in industry. Relatively simple medical instruments measure temperature, blood pressure (sphygmomanometer), or lung capacity (spirometer). More complex instruments include the familiar X-ray machines and electroencephalographs and electrocardiographs, which detect electrical signals generated by the brain and heart, respectively.

Two of the most complex medical instruments now in use are the CAT (computerized axial tomography) and NMR (nuclear magnetic resonance) scanners, which can visualize body parts in three dimensions. The analysis of tissue samples using highly sophisticated methods of chemical analysis is also important in biomedical research.

23.5 BATTERY SILVER OXIDE-ZINC CELL

An alkaline system, this cell features a silver oxide cathode and a powdered zinc anode. Because it will tolerate relatively heavy current load pulses and has a high, nearly constant, 1.5-volt operating voltage, the silver oxide-zinc cell is commonly used in watches, cameras, and hearing aids. In spite of its high cost, the outstanding current-carrying capability of this cell has resulted in its use as military torpedo batteries. Miniature cells can be obtained with either divalent silver oxide or monovalent silver oxide, the former usually having somewhat higher capacity.

Lithium cells are the area of battery technology that has attracted the most research in recent years is a class of cells with a lithium anode. Because of the high chemical activity of lithium, non-aqueous (organic or inorganic) electrolytes have to be used. Such electrolytes include selected solid crystalline salts (see below).

This whole new science has encouraged the commercial production of cells having no space between the anode and the liquid cathode, an unlikely condition for success in aqueous systems. A stable protective layer automatically forms on the lithium but breaks down on discharge to permit high-current operation at nearly constant voltages near 3.6 volts.

By traditional measures, this allows very high power density and energy density. Lithium cells are especially attractive for use in certain aerospace applications, terrestrial portable military equipment, and such civilian applications as personal paging systems, heart pacers, and automated cameras.

Lithium-iron sulfide cells in miniature sizes offer high capacity and low cost for light loads. In operations requiring 1.5 to 1.8 volts, they are a potential substitute for some silver oxide-zinc cells. In constructions where the electrodes consist of rolled up ("jelly roll") strips like those of small nickel-cadmium cells, higher power density is obtained while still retaining high capacity for premium general-purpose use. A typical electrolyte might be lithium tetrafluoborate salt in a solvent mixture of propylene carbonate, 2-methyl-2-oxazolidone, and dimethoxyethane.

Lithium-manganese dioxide cell systems have slowly gained increasingly wider application in small appliances. Cells of this kind have an operating voltage of 2.8 volts each and offer high energy density and relatively low cost compared to some other lithium cell possibilities.

The lithium-carbon mono-fluoride system has been among the more successful early commercial lithium cells. It has been used extensively in cameras and smaller devices, providing about 3.2 volts per cell, high power density, and long shelf life. Good low-temperature performance and a flat voltage-time discharge relationship are provided as well. The cost of carbon mono-fluoride (CF_x) is high, however.

Lithium-thionyl chloride cells provide the highest energy density and power density commercially available. Thionyl chloride serves not only as the electrolyte solvent but also as the cathode material. A runaway reaction between the lithium anode and the adjacent liquid cathode material is prevented by the formation of a film of lithium chloride salt on the lithium.

The electrical contact and reaction centre of the cathode are composed of porous pressed and bonded carbon powder. The performance of this type of cell system at room temperature is very impressive. Moreover, the cell can operate at -54° C, well below the point where aqueous systems function. Because of its high energy density, the lithium-thionyl chloride cell must be used with care and not be burned or

159

disposed of casually. Such cells are useful for powering military equipment, providing backup power for aerospace systems, and operating personal pagers.

Lithium-sulfur dioxide cells have been used extensively for some emergency-aircraft power units and in military cold-weather applications (*e.g.*, radio operation). The cathode consists of a gas under pressure with another chemical as electrolyte salt; this is analogous to the thionyl chloride electrolyte and its liquid cathode. The system functions well but has been found to occasionally vent noxious sulfur dioxide, especially after cold discharge and subsequent warm-up. The release of corrosive or toxic gases by any type of cell in a closed space constitutes a significant design disadvantage.

23.6 TELEVISION, ELECTRONIC SYSTEMS

Mechanical systems lacked sensitivity, as became progressively manifest with attempts to increase the number of lines and thus the degree of definition of the pictures. Swinton and others had pointed out that television pictures, for good quality and definition on a screen of reasonable size, would need to be analyzed into at least 100,000 and preferably 200,000 elements.

Since the number of elements is approximately equal to the square of the number of lines, it can be seen that any system using 30 or even 100 lines would be inadequate--300 being more nearly the minimum. Although mechanical systems were with difficulty made to operate on 200 and more lines, thought increasingly turned toward the greater potential of electronic methods.

A most important landmark was V.K. Zworykin's patent, first filed in 1923, for the iconoscope camera tube. Later, he constructed such a tube; and by 1932 the Radio Corporation of America (RCA), with an improved cathode-ray tube for the receiver, demonstrated all-electronic television (initially on 120 lines), so proving the soundness of Swinton's theoretical ideas.

The compactness and convenience of the electronic camera were remarkable, and its sensitivity, greatly aided by the unique "storage" feature of the iconoscope, was comparable with motion-picture cameras of the time.

Continuing work on electronic systems was greatly stimulated. In the United States the development was mainly carried out in the RCA laboratories; very soon the number of scanning lines was increased to 343, and other improvements followed rapidly. German investigators also were active, especially in the development of high-vacuum cathode-ray tubes. By 1935 a regular broadcasting service had begun in Germany, though with medium definition only--180 lines. In The Netherlands, too, the Philips Laboratories took up television research.

In Great Britain, Electric and Musical Industries (EMI) set up in 1931 a television research group under Isaac Shoenberg. The team produced a complete and practical system, including all of the complex electronics surrounding the camera and receiver tubes, as well as the intervening control and amplifying circuits.

Shoenberg saw the need to establish a system that would endure for many years, since any subsequent changes in basic standards-- particularly the number of scanning lines and their repetition rate-- could give rise to severe technical or economic problems. He therefore proposed the use of 405 lines with 50 frames per second, and interlaced scanning to give 25 pictures per second without flicker.

The government authorized the British Broadcasting Corporation to adopt these standards, as well as the complete EMI system. Initially, and for a short time only, the system was under comparison with alternate broadcasts from a 240-line, 25-picture system developed by the Baird company. The latter employed mechanical scanning methods in the camera and suffered from lack of sensitivity as well as other limitations.

By the mid-1930s electronic television was fast advancing in all its aspects. Important questions were the settling of basic standards (number of lines and frames per second) before the introduction of public broadcasting services in the United States and elsewhere, though these questions were not everywhere fully resolved until about 1951.

The United States soon adopted a picture repetition rate of 30 per second, while in Europe it became 25. These two standards have been perpetuated, and all the countries of the world use one or the other, though technical advances have now obviated the original need for disparity.

The arguments in relation to the number of lines were based on the need for an effective compromise between, on the one hand, adequate picture definition, and, on the other, a frequency bandwidth that could be technically and economically acceptable.

World standardization has never been achieved, though for new television services all countries are adopting one of two main standards, namely, 525 lines per picture at 30 pictures per second--the

161

United States standard--and 625 lines at 25 pictures per second, usually known as the European standard. Complications arise when programs are transmitted between countries using different standards.

Technical advances have been continuous, particularly the great improvements in camera tubes (*e.g.,* the image orthicon and the Vidicon, the latter of which was at last able effectively to exploit the photoconductivity principle) that were made from 1945 onward. By the early 1950s technology had progressed so far, and television had become so widely established, that the time was ripe to tackle in earnest the problem of creating television images in natural colours.

23.7 ROCKET AND GUIDED-MISSILE SYSTEMS

The principal categories of tactical guided missiles are antitank and assault, air-to-surface, air-to-air, anti-ship, and surface-to-air. Distinctions between these categories were not always clear, the launching of both antitank and infantry antiaircraft missiles from helicopters being a case in point.

One of the most important categories of guided missile to emerge after World War II was the antitank, or anti-armour, missile. The guided assault missile, for use against bunkers and structures, was closely related. A logical extension of unguided infantry antitank weapons carrying shaped-charge warheads for penetrating armour, guided antitank missiles acquired considerably more range and power than their shoulder-fired predecessors. While originally intended for issue to infantry formations for self-protection, the tactical flexibility and utility of guided antitank missiles led to their installation on light trucks, on armoured personnel carriers, and, most important, on antitank helicopters.

The first guided antitank missiles were controlled by electronic commands transmitted along extremely thin wires played out from a spool on the rear of the missile. Propelled by solid-fuel sustainer rockets, these missiles used aerodynamic fins for lift and control. Tracking was visual, by means of a flare in the missile's tail, and guidance commands were generated by a hand-operated joystick.

In operating these missiles, the gunner simply superimposed the tracking flare on the target and waited for impact. The missiles were typically designed to be fired from their carrying containers, with the total package small enough to be carried by one or two men. Germany was developing weapons of this kind at the end of World War II and may have fired some in battle.

162

After the war French engineers adapted the German technology and developed the SS-10/SS-11 family of missiles. The SS-11 was adopted by the United States as an interim helicopter-fired antitank missile pending the development of the TOW (for tube-launched, optically tracked, wire-guided) missile. Because it was designed for greater range and hitting power, TOW was mounted primarily on vehicles and, particularly, on attack helicopters.

Helicopter-fired antitank missiles were first used in combat when the U.S. Army deployed several TOW-equipped UH-1 "Hueys" to Vietnam in response to the 1972 communist Easter offensive. TOW was the principal U.S. anti-armour munitions until Hellfire, a more sophisticated helicopter-fired missile with semi-active laser and passive infrared homing, was mounted on the Hughes AH-64 Apache attack helicopter in the 1980s.

The British Swingfire and the French-designed, internationally marketed MILAN (missile d'infanterie léger antichar, or "light infantry antitank missile") and HOT (haut subsonique optiquement téléguidé tiré d'un tube, or "high-subsonic, optically teleguided, tube-fired") were similar in concept and capability to TOW.

The Soviets developed an entire family of antitank guided missiles beginning with the AT-1 Snapper, the AT-2 Swatter, and the AT-3 Sagger. The Sagger, a relatively small missile designed for infantry use on the lines of the original German concept, saw use in Vietnam and was used with conspicuous success by Egyptian infantry in the Suez Canal crossing of the 1973 Arab-Israeli War. The AT-6 Spiral, a Soviet version of TOW and Hellfire, became the principal anti-armour munitions of Soviet attack helicopters.

Many antitank missile systems of later generations transmitted guidance commands by radio rather than by wire, and semi-active laser designation and passive infrared homing also became common. Guidance and control methods were more sophisticated than the original visual tracking and manual commands. TOW, for example, required the gunner simply to centre the reticle of his optical sight on the target, and the missile was tracked and guided automatically. Extremely thin optical fibres began to replace wires as a guidance link in the 1980s.

23.8 TELESCOPE. LIGHT GATHERING AND RESOLUTION

The most important of all the powers of an optical telescope is its light-gathering power. This capacity is strictly a function of the diameter of

163

the clear objective--that is, the aperture--of the telescope. Comparisons of different-sized apertures for their light-gathering power are calculated by the ratio of their diameters squared; for example, a 25-centimetre objective will collect four times the light of a 12.5-centimetre objective [(25 ×25) (12.5 ×12.5)] = 4. The advantage of collecting more light with a larger-aperture telescope is that one can observe fainter stars, nebulas, and very distant galaxies.

Resolving power is another important feature of a telescope. This is the ability of the instrument to distinguish clearly between two points whose angular separation is less than the smallest angle that the observer's eye can resolve.

The resolving power of a telescope can be calculated by the formula resolving power = 11.25 seconds of arc/d,

where d is the diameter of the objective expressed in centimetres. Thus, a 25-centimetre-diameter objective has a theoretical resolution of 0.45 second of arc and a 250-centimetre telescope has one of 0.045 second of arc.

An important application of resolving power is in the observation of visual binary stars. Here, one star is routinely observed as it revolves around a second star. Many observatories conduct extensive visual binary observing programs and publish catalogues of their observational results. One of the major contributors in this field is the United States Naval Observatory in Washington, D.C.

Most refractors currently in use at observatories have equatorial mountings. (The mounting describes the orientation of the physical bearings and structure that permits a telescope to be pointed at a celestial object for viewing.) In the equatorial mounting, the polar axis of the telescope is constructed parallel to the Earth's axis. The polar axis supports the declination axis of the instrument.

Declination is measured on the celestial sky north or south from the celestial equator. The declination axis makes it possible for the telescope to be pointed at various declination angles as the instrument is rotated about the polar axis with respect to right ascension. Right ascension is measured along the celestial equator from the vernal equinox (*i.e.,* the position on the celestial sphere where the Sun crosses the celestial equator from south to north on the first day of spring). Declination and right ascension are the two coordinates that define a celestial object on the celestial sphere.

Declination is analogous to latitude, and right ascension is analogous to longitude. Graduated dials are mounted on the axis to permit the observer to point the telescope precisely. To track an object, the telescope's polar axis is driven smoothly by an electric motor at a sidereal rate--namely, at a rate equal to the rate of rotation of the Earth with respect to the stars.

Thus, one can track or observe with a telescope for long periods of time if the sidereal rate of the motor is very accurate. High-accuracy, motor-driven systems have become readily available with the rapid advancement of quartz-clock technology. Most major observatories now rely on either quartz or atomic clocks to provide accurate sidereal time for observations as well as to drive telescopes at an extremely uniform rate.

A notable example of a refracting telescope is the 66-centimetre refractor of the U.S. Naval Observatory. This instrument was used by the astronomer Asaph Hall to discover the two moons of Mars, Phobos and Deimos, in 1877. Today, the telescope is used primarily for observing double stars. The 91-centimetre refractor at Lick Observatory on Mount Hamilton, Calif., U.S., and the one-metre instrument at Yerkes Observatory in Williams Bay, Wis., U.S., are the largest refracting systems currently in operation (Table 1).

Another type of refracting telescope is the astrograph, which usually has an objective diameter of approximately 20 centimetres. The astrograph has a photographic plate-holder mounted in the focal plane of the objective so that photographs of the celestial sphere can be taken. The photographs are usually taken on glass plates.

The principal application of the astrograph is to determine the positions of a large number of faint stars. These positions are then published in catalogues such as the *AGK*3 and serve as reference points for deep-space photography.

23.9 ACOUSTICS EARLY EXPERIMENTATION

The origin of the science of acoustics is generally attributed to the Greek philosopher Pythagoras (6th century BC), whose experiments on the properties of vibrating strings that produce pleasing musical intervals were of such merit that they led to a tuning system that bears his name.

Aristotle (4th century BC) correctly suggested that a sound wave propagates in air through motion of the air--a hypothesis based more on philosophy than on experimental physics; however, he also incorrectly suggested that high frequencies propagate faster than low frequencies--an error that persisted for many centuries.

165

Vitruvius, a Roman architectural engineer of the 1st century BC, determined the correct mechanism for the transmission of sound waves, and he contributed substantially to the acoustic design of theatres. In the 6th century AD, the Roman philosopher Boethius documented several ideas relating science to music, including a suggestion that the human perception of pitch is related to the physical property of frequency.

The modern study of waves and acoustics is said to have originated with Galileo Galilei (1564-1642), who elevated to the level of science the study of vibrations and the correlation between pitch and frequency of the sound source. His interest in sound was inspired in part by his father, who was a mathematician, musician, and composer of some repute. Following Galileo's foundation work, progress in acoustics came relatively rapidly.

The French mathematician Marin Mersenne studied the vibration of stretched strings; the results of these studies were summarized in the three Mersenne's laws. Mersenne's *Harmonicorum Libri* (1636) provided the basis for modern musical acoustics. Later in the century Robert Hooke, an English physicist, first produced a sound wave of known frequency, using a rotating cog wheel as a measuring device. Further developed in the 19th century by the French physicist Félix Savart, and now commonly called Savart's disk, this device is often used today for demonstrations during physics lectures.

In the late 17th and early 18th centuries, detailed studies of the relationship between frequency and pitch and of waves in stretched strings were carried out by the French physicist Joseph Sauveur, who provided a legacy of acoustic terms used to this day and first suggested the name acoustics for the study of sound.

One of the most interesting controversies in the history of acoustics involves the famous and often misinterpreted "bell-in-vacuum" experiment, which has become a staple of contemporary physics lecture demonstrations. In this experiment the air is pumped out of a jar in which a ringing bell is located; as air is pumped out, the sound of the bell diminishes until it becomes inaudible.

As late as the 17th century many philosophers and scientists believed that sound propagated via invisible particles originating at the source of the sound and moving through space to affect the ear of the observer. The concept of sound as a wave directly challenged this view, but it was not established experimentally until the first bell-in-vacuum experiment was performed by Athanasius Kircher, a German scholar, who described it in his book *Musurgia Universalis* (1650). Even after pumping the air out of the jar, Kircher could still hear the bell, so he concluded incorrectly that air was not required to transmit sound.

In fact, Kircher's jar was not entirely free of air, probably because of inadequacy in his vacuum pump. By 1660 the Anglo-Irish scientist Robert Boyle had improved vacuum technology to the point where he could observe sound intensity decreasing virtually to zero as air was pumped out. Boyle then came to the correct conclusion that a medium such as air is required for transmission of sound waves. Although this conclusion is correct, as an explanation for the results of the bell-in-vacuum experiment it is misleading.

Even with the mechanical pumps of today, the amount of air remaining in a vacuum jar is more than sufficient to transmit a sound wave. The real reason for a decrease in sound level upon pumping air out of the jar is that the bell is unable to transmit the sound vibrations efficiently to the less dense air remaining, and that air is likewise unable to transmit the sound efficiently to the glass jar.

Thus, the real problem is one of an impedance mismatch between the air and the denser solid materials--and not the lack of a medium such as air, as is generally presented in textbooks. Nevertheless, despite the confusion regarding this experiment, it did aid in establishing sound as a wave rather than as particles.

23.10 EXPLOITING RENEWABLE ENERGY SOURCES

Growing concern over the world's ever-increasing energy needs and the prospect of rapidly dwindling reserves of oil, natural gas, and uranium fuel have prompted efforts to develop viable alternative energy sources. The volatility and uncertainty of the petroleum fuel supply were dramatically brought to the fore during the energy crisis of the 1970s caused by the abrupt curtailment of oil shipments from the Middle East to many of the highly industrialized nations of the world. It also has been recognized that the heavy reliance on fossil fuels has had an adverse impact on the environment.

Gasoline engines and steam-turbine power plants that burn coal or natural gas emit substantial amounts of sulfur dioxide and nitrogen oxides into the atmosphere. When these gases combine with atmospheric water vapour, they form sulfuric acid and nitric acids, giving rise to highly acidic precipitation. The combustion of fossil fuels also releases carbon dioxide. The amount of this gas in the atmosphere has steadily risen since the mid-1800s largely as a result of the growing consumption of coal, oil, and natural gas.

More and more scientists believe that the atmospheric build-up of carbon dioxide (along with that of other industrial gases such as methane and chlorofluorocarbons) may induce a greenhouse effect, raising the surface temperature of the Earth by increasing the amount of heat trapped in the lower atmosphere. This condition could bring about climatic changes with serious repercussions for natural and agricultural ecosystems.

Many countries have initiated programs to develop renewable energy technologies that would enable them to reduce fossil-fuel consumption and its attendant problems. Fusion devices are believed to be the best long-term option, since their primary energy source would be the hydrogen isotope deuterium abundantly present in ordinary water. Other technologies that are being actively pursued are those designed to make wider and more efficient use of the energy in sunlight, wind, moving water, and terrestrial heat (*i.e.,* geothermal energy). The amount of energy in such renewable and virtually pollution-free sources is large in relation to world energy needs, yet at the present time only a small portion of it can be converted to electric power at reasonable cost.

A variety of devices and systems has been created to better tap the energy in sunlight. Among the most efficient are photovoltaic systems that transform radiant energy from the Sun directly into electricity by means of silicon or gallium arsenide solar cells. Large arrays consisting of thousands of these semiconductor cells can function as central power stations. Other systems, which are still under development, are designed to concentrate solar radiation not only to generate electric power but also to produce high-temperature process heat for various applications. These systems employ a number of different components, including large parabolic concentrators and heat engines of the Stirling engine type (see above). Another approach involves the use of flat-plate solar collectors to provide space heating for commercial and residential buildings.

Although wind is intermittent and diffuse, it contains tremendous amounts of energy. Sophisticated wind turbines have been developed to convert this energy to electric power. The utilization of wind energy systems grew discernibly during the 1980s. For example, more than 15,000 wind turbines are now in operation in Hawaii and California at specially selected sites. Their combined power rating of 1,500 megawatts is roughly equal to that of a conventional steam-turbine power installation.

Converting the energy in moving water to electricity has been a long-standing technology. Yet, hydroelectric power plants are estimated to provide only about 2 percent of the world's energy requirements. The technology involved is simple enough: hydraulic turbines change the energy of fast-flowing or falling water into mechanical energy that drives power generators, which produce electricity. Hydroelectric power plants, however, generally require the building of costly dams. Another factor that limits any significant increase in hydroelectric power production is the scarcity of suitable sites for additional installations except in certain regions of the world.

In certain coastal areas of the world, as, for example, the Rance River estuary in Brittany, Fr., hydraulic turbine-generator units have been used to harness the great amount of energy in ocean tides. At most such sites, the capital costs of constructing damlike structures with which to trap and store water are prohibitive, however.

Geothermal energy flows from the hot interior of the Earth to the surface in steam or hot water most often in areas of active volcanism. Geothermal reservoirs with temperatures of 180° C or higher are suitable for power generation. The earliest commercial geothermal power plant was built in 1904 in Larderello, Italy. Today, steam from wells drilled to depths of hundreds of metres drives the plant's turbine generators to produce about 190 megawatts of electricity.

Geothermal plants have been built in a number of other countries, including El Salvador, Japan, Mexico, New Zealand, and the United States. The principal U.S. plant, located at The Geysers north of San Francisco, can generate up to 1,900 megawatts, though production may be restricted to prolong the life of the steam field.

23.11 MAGNETO-HYDRODYNAMIC

Magneto-hydrodynamic power generator (MHD) is one of a class of devices that generate electric current by means of the interaction of an electrically conducting fluid and a magnetic field. Various countries--including Japan, China, Poland, Russia, and the United States--have undertaken active developmental programs in magneto-hydrodynamic technology, since MHD power plants offer the potential for large-scale electrical power generation at reasonable cost with comparatively little detrimental impact on the environment.

Generators of the MHD type are also attractive for the production of large electrical power pulses, and their first practical application has been for this kind of service (see below).

The underlying principle of MHD power generation is elegantly simple. An electrically conducting fluid is driven by a primary energy source (*e.g.,* combustion of coal or a gas) through a magnetic field, resulting in the establishment of an electromotive force within the conductor in accordance with the principle established by Faraday.

Furthermore, if the conductor is an electrically conducting gas, it will expand, and so the MHD system constitutes a heat engine involving an expansion from high to low pressure in a manner similar to that of a gas turbine. The MHD system, however, involves a volume interaction between a gas and the magnetic field through which it is passing (see below), whereas the gas turbine operates through the gas interaction with the surfaces of a rotating blade system. It is, in effect, a system that depends on volume rather than surface interaction.

The MHD generator can properly be viewed as an electromagnetic turbine because its output is obtained from the conducting gas-magnetic field interaction directly in electrical form rather than in mechanical form, as in the case of a gas (or steam) turbine.

Electrical conduction in gases occurs when electrons are available to be organized into an electric current in response to an applied or induced electric field. The electrons may be either injected or generated internally, and, because of the electrostatic forces involved, they require the presence of corresponding positive charge from ions to maintain electrical neutrality. An electrically conducting gas consists in general of electrons, ions to balance the electric charge, and neutral atoms or molecules. Such a gas is termed a plasma.

In MHD generators, electrons for supporting the flow of current can be obtained in either of two ways: by heating the gas to a sufficiently high temperature to yield electrons through ionization or by the induction of a sufficiently strong electric field in a manner similar to that in gas-discharge devices.

These methods are referred to as thermal ionization and non-equilibrium ionization, respectively. In either case, the mechanism of energy transfer from the flowing fluid to the electrical output can be thought of as a coupling of the electron-comprised gas to the ions through electromagnetic forces; the ions in turn are embedded in the background of atomic or molecular gas and lack mobility by virtue of their being coupled to the molecules or ions through collision processes described by kinetic behaviour.

Interest in MHD-power generation was originally stimulated by the observation that the interaction of a plasma with a magnetic field could occur at much higher temperatures than were possible in a system consisting of a rotating mechanical system. The limiting performance from the point of view of efficiency in heat engines is established by the Carnot efficiency, obtained from the difference between the absolute hot source temperature, T_1, and the cold sink temperature, T_0, divided by T_1.

For example, when the source temperature is 2,310 K and the sink temperature is that of the environment (say, 294 K), the Carnot efficiency is slightly less than 90 percent. Allowing for the inefficiencies introduced by finite heat transfer rates and component inefficiencies in real heat engines, a system employing an MHD generator offers the potential of an ultimate efficiency in the range 60 to 65 percent.

This is to be compared with 35 to 36 percent achieved by a modern coal-fired, steam-turbine plant with scrubbers (devices that absorb sulfur dioxide from exhaust gases); 40 percent with a natural gas-fired, steam-turbine plant; and about 46 percent projected for gas-fired, combined gas-steam turbine installations. The implications of this efficiency improvement are an enhanced utilization of primary fuel resources due to higher thermodynamic efficiency and a lower emission of environmental pollutants.

23.12 FERROUS METALS

From 1500 to the 20th century, metallurgical development was still largely concerned with improved technology in the manufacture of iron and steel. In England, the gradual exhaustion of timber led first to prohibitions on cutting of wood for charcoal and eventually to the introduction of coke, derived from coal, as a more efficient fuel.

Thereafter the iron industry expanded rapidly in Great Britain, which became the greatest iron producer in the world. The crucible process for making steel, introduced in England in 1740, by which bar iron and added materials were placed in clay crucibles heated by coke fires, resulted in the first reliable steel made by a melting process.

One difficulty with the bloomery process for the production of soft bar iron was that, unless the temperature was kept low (and the output therefore small), it was difficult to keep the carbon content low enough so that the metal remained ductile. This difficulty was overcome by melting high-carbon pig iron from the blast furnace in the puddling process, invented in Great Britain in 1784.

In it, melting was accomplished by drawing hot gases over a charge of pig iron and iron ore held on the furnace hearth. During its manufacture the product was stirred with iron rabbles (rakes), and, as it became pasty with loss of carbon, it was worked into balls, which were subsequently forged or rolled to a useful shape.

The product, which came to be known as wrought iron, was low in elements that contributed to the brittleness of pig iron and contained enmeshed slag particles that became elongated fibres when the metal was forged. Later, the use of a rolling mill equipped with grooved rolls to make wrought-iron bars was introduced.

The most important development of the 19th century was the large-scale production of cheap steel. Prior to about 1850, the production of wrought iron by puddling and of steel by crucible melting had been conducted in small-scale units without significant mechanization. The first change was the development of the open-hearth furnace by William and Friedrich Siemens in Britain and by Pierre and Émile Martin in France.

Employing the regenerative principle, in which outgoing combusted gases are used to heat the next cycle of fuel gas and air, this enabled high temperatures to be achieved while saving on fuel. Pig iron could then be taken through to molten iron or low-carbon steel without solidification, scrap could be added and melted and iron ore could be melted into the slag above the metal to give a relatively rapid oxidation of carbon and silicon--all on a much enlarged scale.

Another major advance was Henry Bessemer's process, patented in 1855 and first operated in 1856, in which air was blown through molten pig iron from tuyeres set into the bottom of a pear-shaped vessel called a converter.

Heat released by the oxidation of dissolved silicon, manganese, and carbon was enough to raise the temperature above the melting point of the refined metal (which rose as the carbon content was lowered) and thereby maintain it in the liquid state. Very soon Bessemer had tilting converters producing 5 tons in a heat of one hour, compared with four to six hours for 50 kilograms (110 pounds) of crucible steel and two hours for 250 kilograms of puddled iron.

Neither the open-hearth furnace nor the Bessemer converter could remove phosphorus from the metal, so that low-phosphorus raw materials had to be used. This restricted their use from areas where phosphoric ores, such as those of the Minette range in Lorraine, were a main European source of iron.

The problem was solved by Sidney Gilchrist Thomas, who demonstrated in 1876 that a basic furnace lining consisting of calcined dolomite, instead of an acidic lining of siliceous materials, made it

172

possible to use a high-lime slag to dissolve the phosphates formed by the oxidation of phosphorus in the pig iron. This principle was eventually applied to both open-hearth furnaces and Bessemer converters.

As steel was now available at a fraction of its former cost, it saw an enormously increased use for engineering and construction. Soon after the end of the century it replaced wrought iron in virtually every field. Then, with the availability of electric power, electric-arc furnaces were introduced for making special and high-alloy steels.

The next significant stage was the introduction of cheap oxygen, made possible by the invention of the Linde-Frankel cycle for the liquefaction and fractional distillation of air. The Linz-Donawitz process, invented in Austria shortly after World War II, used oxygen supplied as a gas from a tonnage oxygen plant, blowing it at supersonic velocity into the top of the molten iron in a converter vessel. As the ultimate development of the Bessemer/Thomas process, oxygen blowing became universally employed in bulk steel production.

24. SOCIAL AMENITIES, NEEDS AND DEMANDS

24.1 AUTOMOBILE FUTURE SYSTEMS

Expansion of the total potential automotive market in the future and concern for the environment may be expected to change cars of the future. Special-purpose vehicles designed for specific urban or rural functions, with appropriate power systems for each type of use, may be needed. Possibilities include electric, solar, steam, gas-turbine, hybrid combinations, and other power sources.

Modern electric cars and trucks were manufactured in small numbers in Europe, Japan, and the United States beginning in the 1980s. However, electric propulsion is only possible for relatively short-range vehicles, using power from batteries or fuel cells. In a typical system, a group of lead-acid batteries, connected in a series, power electric AC induction motors to propel the vehicle.

A solid-state rectifier, or power inverter, changes the direct current supplied by the battery pack to an alternating current output that is controlled by the driver using an accelerator pedal to vary the output voltage. Because of the torque characteristics of electric motors, conventional gear-type transmissions are not needed in most designs. Weight and drag reduction, as well as regenerative systems to recover energy that would otherwise be lost, are important considerations in extending battery life.

Conventional storage-battery systems do not have high power-to-weight ratios for acceleration or energy-to-weight ratios for driving range to match gasoline-powered general-purpose vehicles. Special-purpose applications, however, may be practical because of the excellent low-emission characteristics of the system. Such systems have been used to power vehicles on the Moon. Storage-battery systems also may be coupled with gasoline or other power sources to form a hybrid power plant with electric operation in cities and gasoline operation in less congested areas.

Renewed interest in two-stroke cycle gasoline engines has led several firms in the early 1990s to develop designs related to patents of the Orbital Engine Company of Australia. Air-assisted direct injection of fuel permits very lean-burning stratified combustion. A variable exhaust port confines exhaust gas within the cylinders. Electronic controls provide for proper actuation under varying speeds and loads to produce lower emissions and higher fuel economy with improved power-to-weight characteristics.

Compressed natural gas (CNG) and blends derived from methanol (wood alcohol) and ethanol (grain alcohol) are being studied as fuels for the future because they may be produced from readily available biomass sources and have potential for lean burning, high efficiency, and lower emissions. Their problems relate to the large pressurized fuel tanks required because of the lower energy density of the fuel, their poor cold-starting characteristics, and, in the case of alcohols, the corrosive character of the fuel.

Steam power plants have been re-examined in the light of modern technology and new materials. The continuous-combustion process used to heat the steam generator offers potentially improved emission characteristics.

Gas turbines have been tested extensively and have good torque characteristics, operate on a wide variety of fuels, have high power-to-weight ratios, meet emission standards, and offer quiet operation. Some studies have shown that the advantages of the system are best realized in heavy-duty vehicles operating on long, nearly constant speed runs.

Efficiencies and operating characteristics can be improved by increasing operating temperatures. This may be commercially feasible utilizing ceramic materials that are cost-effective. Successful designs require regenerative systems to recover energy from hot exhaust gas and transfer it to incoming air. This improves fuel economy, reduces exhaust temperatures to safer levels, and eliminates the need for a muffler in some designs.

Nuclear energy offers the advantage of extremely low fuel weight. The obstacle for automotive use, however, is the great weight and volume of shielding required to protect the occupants from excessive nuclear radiation.

A number of other energy-conversion systems have been studied for automotive applications. Many of these are variations of engine combustion cycles such as turbocharged gasoline and diesel (two- and four-stroke) designs. Free-piston and other geometric configurations, including rotary engines, offer other possibilities. One Japanese company was manufacturing automobiles driven by Wankel-type rotary engines in 1992.

The final third of the 20th century saw unprecedented interaction of the federal governments of most industrial nations in the automotive

175

design process. The result was a 100-percent improvement in average fuel economy and the reduction of various polluting by-products of automobile operation.

These outstanding technical advancements were not made without economic consequences. According to a study by the Motor Vehicle Manufacturers Association of the United States, Inc., the average consumer expenditure for new cars increased 83 percent between 1967 and 1990 because of added safety and emission control features. This is in addition to the costs associated with fuel economy improvement, which may be offset by reduced fuel purchases. General economic inflation during the period was removed from the calculation.

The advent during the 1990s of regulations requiring "zero emissions" from some vehicles in certain areas rekindled world interest in new battery technology. Battery systems that offer higher energy density became the subject of joint research by federal and auto industry scientists. Non-commercial solar-powered electric demonstration vehicles were built by universities and manufacturers. Solar-collector areas proved to be too large for conventional cars, however. Development continued on cell design and motor-vehicle-component power supply requirements such as heater or air-conditioning fans.

The modern automobile is responsible for about one-half of the rubber, one-third of the platinum, one-sixth of the aluminum, one-seventh of the steel, and one-tenth of the copper consumed in the United States each year. Because the automobile is likely to remain an important part of the transportation system, it requires continuing improvement in safety and emission control as well as in its older values of performance and costs.

24.2 RAILROAD OPERATIONS

The overhead costs of COFC and TOFC are considerable. Both require terminals with high-capacity trans-shipment cranage and considerable space for internal traffic movement and storage. TOFC also has a cost penalty in the deadweight of the highway trailers' running gear that has to be included in a TOFC train's payload. Two principal courses have been taken by railroads to improve the economics of their inter-modal operations.

One is to limit their trans-shipment terminals to strategically located and well-equipped hubs, from which highway collection and delivery services radiate over longer distances; as a result, the railroad can carry the greater part of its inter-modal traffic in full terminal-to-terminal trainloads, or unit trains.

The other course has been to minimize the tare weight of rail inter-modal vehicles by such techniques as skeletal frame construction and, as in the double-stack COFC units described above, articulation of car frames over a single truck. Even so, North American railroads have not been able to make competitively priced TOFC remunerative unless the rail component of the transit is more than about 600 miles.

Two different managerial approaches to inter-modal freight service have developed in the United States. Some of the major railroads have organized to manage and market complete door-to-door transits themselves; others prefer simply to wholesale inter-modal train space to third parties. These third parties organize, manage, and bill the whole door-to-door transit for an individual consignor.

Given the shorter intercity distances, European railroads have found it more difficult to operate viable TOFC services. The loading of a highway box trailer on a railcar of normal frame height without infringing European railroads' reduced vertical clearances was solved by French National Railways in the 1950s.

The answer was a railcar with floor pockets into which the trailer's wheels could be slotted, so that the trailer's floor ended up parallel with that of the railcar. Even so, there were limitations on the acceptable height of box trailers.

Other railroads were prompted to begin TOFC in the 1960s when the availability of heavy tonnage cranes at new container terminals simplified the placing of trailers in the so-called "pocket" cars. Initial TOFC service development was primarily over long and mostly international trade routes, such as from The Netherlands, Belgium, and northern Germany to southern Germany, Austria, and Italy.

In 1978 the West German government decided to step up investment in its railways for environmental and energy-saving reasons. Its plans included a considerable subsidy of railroad inter-modal operation, including TOFC. Similar support of inter-modal development, for the same reasons, was subsequently provided for their national railways by the Austrian and Swiss governments.

The German railroad (and also Scandinavian railroads) has more generous vertical clearances than the European norm. Whereas other European mainland railroads, even with pocket cars, can only operate TOFC over a few key trunk routes, the German Federal Railway could

177

use the financial support to launch TOFC as well as COFC service between most of its major production and consumption areas.

The Germans, followed by the Austrians and Swiss, also developed a particularly costly inter-modal technology. They call it "Rolling Highway" (Rollende Landstrasse), because it employs low-floor cars that, coupled into a train, form an uninterrupted drive-on, drive-off roadway for highway trucks or tractor-trailer rigs. Rolling Highway cars are carried on four- or six-axle trucks with wheels of only 14-inch diameter so as to lower their floors sufficiently to secure the extra vertical clearance for highway vehicles loaded without their wheels pocketed.

Platforms bridge the gap between the close-coupled railcars. To allow highway vehicles to drive on or off the train yet enable a locomotive to couple to it without difficulty, the train-end low-floor cars have normal-height draft-gear headstocks that are hinged and can be swung aside to open up the train's roadway. Truck drivers travel in a passenger car added to the train.

In the face of growing trade between north-western and south-eastern Europe, Austria and Switzerland have imposed restraints on use of their countries as a transit corridor by over-the-highway freight to safeguard their environments. Consequently, the highest inter-modal traffic growth rates in Europe were registered by their national railroads in the 1980s and early '90s; and Rolling Highway services proliferated on their transalpine routes.

Primarily to provide for increase in inter-modal traffic, and in particular Rolling Highway trains, the Swiss parliament in 1991 approved a government plan to bore new rail tunnels on each of its key north-south transalpine routes, the Gotthard and the Lötschberg. The new tunnels will be much longer than those at the summit of the existing routes; thus their tracks will be free of the present routes' steep gradients and sharp curves on either side of their tunnels.

24.3 HEATING AND COOLING SYSTEMS

Steam and hot-water heating systems of the late 19th century provided a reasonable means for winter heating, but no practical methods existed for artificial cooling, ventilating, or humidity control. In the forced-air system of heating, air replaced steam or water as the fluid medium of heat transfer, but this was dependent on the development of powered fans to move the air.

Although large, crude fans for industrial applications in the ventilation of ships and mines had appeared by the 1860s, and the Johns Hopkins Hospital in Baltimore had a successful steam-powered forced-air system installed in 1873, the widespread application of this system to buildings only followed the development of electric-powered fans in the 1890s.

Important innovations in cooling technology followed. The development of refrigeration machines for food storage played a role, but the key element was Willis Carrier's 1906 patent that solved the problem of humidity removal by condensing the water vapour on droplets of cold water sprayed into an air-stream. Starting with humidity control in tobacco and textile factories, Carrier slowly developed his system of "man-made weather," finally applying it together with heating, cooling, and control devices as a complete system in Graumann's Metropolitan Theater, Los Angeles, in 1922.

The first office building air-conditioned by Carrier was the 21-story Milam Building (1928) in San Antonio, Texas. It had a central refrigeration plant in the basement that supplied cold water to small air-handling units on every other floor; these supplied conditioned air to each office space through ducts in the ceiling; the air was returned through grills in doors to the corridors and then back to the air-handling units.

A somewhat different system was adopted by Carrier for the 32-story Philadelphia Savings Fund Society Building (1932). The central air-handling units were placed with the refrigeration plant on the 20th floor, and conditioned air was distributed through vertical ducts to the occupied floors and horizontally to each room and returned through the corridors to vertical exhaust ducts that carried it back to the central plant. Both systems of air handling, local and central, are still used in high-rise buildings.

The Great Depression and World War II reduced the demand for air-conditioning systems, and it was not until the building of the United Nations Secretariat in New York City in 1949 that Carrier produced a method of air conditioning that could deal effectively with the large heat loads imposed by the building's all-glass curtain walls.

The conditioned air was delivered not only from the ceiling but also through pipe coil convector units just inside the glass wall. The pipe coil convectors contained centrally supplied warm or cold water to further temper the heat loss or gain at the perimeter; conditioned air and water were centrally supplied from four mechanical floors spaced within the building's 39-story height.

Andreas Sofroniou

Carrier's "Weathermaster" system was energy-intensive, appropriate to the declining energy costs of the time, and it was adopted for most of the all-glass skyscrapers that followed in the next 25 years. In the 1960s the so-called dual-duct system appeared; both warm and cold air were centrally supplied to every part of the building and combined in mixing boxes to provide the appropriate atmosphere.

The dual-duct system also consumed much energy, and, when energy prices began to rise in the 1970s, both it and the Weathermaster system were supplanted by the variable air volume (VAV) system, which supplies conditioned air at a single temperature, the volume varying according to the heat loss or gain in the occupied spaces. The VAV system requires much less energy and is widely used.

In the early 1950s, air-conditioning systems were reduced to very small electric-powered units capable of cooling single rooms. These were usually mounted in windows to take in fresh air and to remove heat to the atmosphere. These units found widespread application in the retrofitting of existing buildings--particularly houses and apartment buildings--and have since found considerable application in new residential buildings.

The relatively high energy costs of the 1970s also prompted interest in various forms of solar heating, both for interior spaces and for domestic hot water, but, except for residential passive solar heating, the relative decline in energy prices in the 1980s made such systems unattractive.

The study of thermodynamics in the late 19th century included the heat-transfer properties of materials and led to the concept of thermal insulation--that is, a material that has a relatively low rate of heat transfer. As building atmospheres became more carefully controlled after 1900, more attention was given to the thermal insulation of building enclosures (envelopes).

One of the best insulators is air, and materials that trap air in small units have low heat-transfer rates; wool and foam are excellent examples. The first commercial insulations, in the 1920s, were mineral wools and vegetable-fibreboards; fibreglass wool appeared in 1938. Foam glass, the first rigid insulating foam, was marketed in the 1930s, and after 1945 a wide variety of plastic foam insulations was developed.

Since the 1970s most building codes have set minimum requirements for insulation of building envelopes, and these have proved to be very cost-effective in saving energy.

24.4 UNDERGROUND, TUBE, OR MÉTRO

Subway underground railway system used to transport large numbers of passengers within urban and suburban areas. Subways are usually built under city streets for ease of construction, but they may take shortcuts and sometimes must pass under rivers. Outlying sections of the system usually emerge aboveground, becoming conventional railways or elevated transit lines. Subway trains are usually made up of a number of cars operated on the multiple-unit system.

The first subway system was proposed for London by Charles Pearson, a city solicitor, as part of a city-improvement plan shortly after the opening of the Thames Tunnel in 1843. After 10 years of discussion, Parliament authorized the construction of 3.75 miles (6 km) of underground railway between Farringdon Street and Bishop's Road, Paddington.

Work on the Metropolitan Railway began in 1860 by cut-and-cover methods--that is, by making trenches along the streets, giving them brick sides, providing girders or a brick arch for the roof, and then restoring the roadway on top. On Jan. 10, 1863, the line was opened using steam locomotives that burned coke and, later, coal; despite sulfurous fumes, the line was a success from its opening, carrying 9,500,000 passengers in the first year of its existence. In 1866 the City of London and Southwark Subway Company (later the City and South London Railway) began work on their "tube" line, using a tunnelling shield developed by J.H. Greathead.

The tunnels were driven at a depth sufficient to avoid interference with building foundations or public-utility works, and there was no disruption of street traffic. The original plan called for cable operation, but electric traction was substituted before the line was opened. Operation began on this first electric underground railway in 1890 with a uniform fare of twopence for any journey on the 3-mile (5-kilometre) line.

In 1900 Charles Tyson Yerkes, an American railway magnate, arrived in London, and he was subsequently responsible for the construction of more tube railways and for the electrification of the cut-and-cover lines. During World Wars I and II the tube stations performed the unplanned function of air-raid shelters.

Many other cities followed London's lead. In Budapest, a 2.5-mile (4-kilometre) electric subway was opened in 1896, using single cars with trolley poles; it was the first subway on the European continent.

Considerable savings were achieved in its construction over earlier cut-and-cover methods by using a flat roof with steel beams instead of a brick arch, and therefore, a shallower trench.

In Paris, the Métro (Chemin de Fer Métropolitain de Paris) was started in 1898, and the first 6.25 miles (10 km) were opened in 1900. The rapid progress was attributed to the wide streets overhead and the modification of the cut-and-cover method devised by the French engineer Fulgence Bienvenue.

Vertical shafts were sunk at intervals along the route; and, from there, side trenches were dug and masonry foundations to support wooden shuttering were placed immediately under the road surfaces. Construction of the roof arch then proceeded with relatively little disturbance to street traffic. This method, while it is still used in Paris, has not been widely copied in subway construction elsewhere.

In the United States the first practical subway line was constructed in Boston between 1895 and 1897. It was 1.5 miles (2.4 km) long and at first used trolley streetcars, or tramcars. Later, Boston acquired conventional subway trains.

New York City opened the first section of what was to become the largest system in the world on Oct. 27, 1904. In Philadelphia, a subway system was opened in 1907, and Chicago's system opened in 1943. Moscow constructed its original system in the 1930s.

In Canada, Toronto opened a subway in 1954; a second system was constructed in Montreal during the 1960s using Paris-type rubber-tired cars. In Mexico City the first stage of a combined underground and surface metro system (designed after the Paris Métro) was opened in 1969. In South America, the Buenos Aires subway opened in 1913. In Japan, the Tokyo subway opened in 1927, the Kyoto in 1931, the Osaka in 1933, and the Nagoya in 1957.

Automatic trains, designed, built, and operated using aerospace and computer technology, have been developed in a few metropolitan areas, including a section of the London subway system, the Victoria Line (completed 1971). The first rapid-transit system to be designed for completely automatic operation is BART (Bay Area Rapid Transit) in the San Francisco Bay area, completed in 1976.

Trains are operated by remote control, requiring only one crewman per train to stand by in case of computer failure. The Washington, D.C., Metro, with an automatic railway control system and 600-foot- (183-metre-) long underground coffered-vault stations, opened its first subway line in 1976.

Air-conditioned trains with lightweight aluminum cars, smoother and faster rides due to refinements in track construction and car-support systems, and attention to the architectural appearance of and passenger safety in underground stations are other features of modern subway construction.

24.5 TRANSISTOR: THE SOLID-STATE REVOLUTION

The invention of the transistor in 1947 by John Bardeen, Walter H. Brattain, and William B. Shockley of the Bell research staff provided the first of a series of new devices with remarkable potential for expanding the utility of electronic equipment.

Transistors, along with such subsequent developments as integrated circuits, are made of crystalline solid materials called semiconductors, which have electrical properties that can be varied over an extremely wide range by the addition of minuscule quantities of other elements.

The electric current in semiconductors is carried by electrons, which have a negative charge, and also by holes, analogous entities that carry a positive charge. The availability of two kinds of charge carriers in semiconductors is a valuable property exploited in many electronic devices made of such materials.

Early transistors were produced using germanium as the semiconductor material, because methods of purifying it to the required degree had been developed during and shortly after World War II. Because the electrical properties of semiconductors are extremely sensitive to the slightest trace of certain other elements, only about one part per billion of such elements can be tolerated in material to be used for making semiconductor devices.

During the late 1950s, research on the purification of silicon succeeded in producing material suitable for semiconductor devices, and new devices made of silicon were manufactured from about 1960. Silicon quickly became the preferred raw material, because it is much more abundant than germanium and thus intrinsically less expensive.

In addition, silicon retains its semiconducting properties at higher temperatures than does germanium. Silicon diodes can be operated at temperatures up to 200° C (392° F), whereas germanium diodes cannot be operated above 85° C. There was one other important property of silicon, not appreciated at the time but crucial to the development of low-cost transistors and integrated circuits: silicon, unlike germanium, forms a tenaciously adhering oxide film with excellent electrical insulating properties when it is heated to high temperatures in the presence of oxygen.

This film is utilized as a mask to permit the desired impurities that modify the electrical properties of silicon to be introduced into it during manufacture of semiconductor devices. The mask pattern, formed by a photolithographic process, permits the creation of tiny transistors and other electronic components in the silicon.

By 1960 vacuum tubes were rapidly being supplanted by transistors, because the latter had become less expensive, did not burn out in service, and were much smaller and more reliable. Computers employed hundreds of thousands of transistors each.

This fact, together with the need for compact, lightweight electronic missile guidance systems, led to the invention of the integrated circuit (IC) independently by Jack Kilby of Texas Instruments Incorporated in 1958 and by Jean Hoerni and Robert Noyce of Fairchild Semiconductor Corporation in 1959. Kilby is usually credited with having developed the concept of integrating device and circuit elements onto a single silicon chip, while Noyce is given credit for having conceived the method for integrating the separate elements.

Early ICs contained about 10 individual components on a silicon chip 3 millimetres (0.12 inch) square. By 1970 the number was up to 1,000 on a chip the same size at no increase in cost. Late in the following year the first microprocessor was introduced. The device contained all the arithmetic, logic, and control circuitry required to perform the functions of a computer's central processing unit. This type of large-scale IC was developed by a team at Intel Corporation, the same company that also introduced the memory integrated circuit in 1971. The stage was now set for the computerization of small electronic equipment.

Until the microprocessor appeared on the scene, computers were essentially discrete pieces of equipment used primarily for data processing and scientific calculations. They ranged in size from minicomputers, comparable in dimensions to a microwave oven, to mainframe systems that took up enough space to fill a large room.

The microprocessor enabled computer engineers to develop microcomputers--systems about the size of a lunch box or smaller but with enough computing power to perform many kinds of business, industrial, and scientific tasks. Such systems made it possible to control a host of small instruments or devices (*e.g.*, numerically controlled lathes and one-armed robotic devices for spot welding) by using standard components programmed to do a specific job. The very existence of computer hardware inside such devices is not apparent to the user.

The large demand for microprocessors generated by these initial applications led to high-volume production and a dramatic reduction in cost. This in turn promoted the use of the devices in many other applications--as, for example, in household appliances and automobiles, for which electronic controls had previously been too expensive to consider. Continued advances in IC technology gave rise to very-large-scale integration (VLSI), which substantially increased the circuit density of microprocessors.

This technological advance, coupled with further cost reductions stemming from improved manufacturing methods, made feasible the mass production of personal computers for use in schools and homes, as well as in offices.

By the mid-1980s, inexpensive microprocessors had stimulated computerization of an enormous variety of consumer products. Common examples included programmable microwave ovens and thermostats, clothes washers and dryers, self-tuning television sets and self-focusing cameras, videocassette recorders and video games, telephones and answering machines, musical instruments, watches, and security systems. Microelectronics also came to the fore in business, industry, government, and other sectors. Microprocessor-based equipment proliferated, ranging from automatic teller machines and point-of-sale terminals in retail stores to automated factory assembly systems and office workstations.

By mid-1986, memory ICs with a capacity of 262,144 bits (binary digits) was available. Within a short time, circuits of this kind having four times that capacity were being produced. By the mid-1990s, microprocessors capable of handling 32-bit words were common, and 64-bit versions were available.

The larger memories and microprocessors contained more than 20 million transistors on a silicon chip less than two centimetres square. In addition, literally tens of thousands of other kinds of ICs for various applications were available, varying in complexity from a few dozen transistors upward.

24.6 ARCHITECTURE, AFTER WORLD WAR II

Initially, the leading interwar architects of modernism, Gropius, Mies van der Rohe, Le Corbusier, Wright, and Aalto, continued to dominate the scene. In the United States, Gropius, with Breuer, introduced

185

modern houses to Lincoln, Mass., a Boston suburb, and formed a group, The Architects Collaborative, the members of which designed the thoroughly modern Harvard Graduate Center (1949-50). Mies became dean of the department of architecture at the Illinois Institute of Technology at Chicago in 1938 and designed its new campus. Crown Hall (1952-56) marked the apogee of this quarter-century project.

Not all the immigrants remained in the United States. Aalto, whose work first appeared on the American scene in the Finnish pavilion at the New York World's Fair and again in the Massachusetts Institute of Technology's Baker Dormitory (1947-49), returned to Finland. The European who might have contributed most was Le Corbusier.

The United Nations buildings at New York City, for which he was a member of a 10-man commission headed by New York architect Wallace Harrison, is a token of the new forms he might have suggested for American cities. His plan for rebuilding Saint-Dié, Fr. (1945), was the inspiration for many city-planning proposals made after mid-century.

Beginning with private houses by Hood, Lescaze, Edward Stone, Neutra, Gropius, and Breuer during the 1930s, American Modernism gradually supplanted the historical styles in a range of building types, including schools and churches; for example, Eliel Saarinen's simple, brick Christ Lutheran Church (1949-50) at Minneapolis, Minn.

After World War II, big industry turned to Modern architects for distinctive emblems of prestige. The Connecticut General Life Insurance Company hired one of the largest modern firms, Skidmore, Owings, and Merrill, to design their new decentralized headquarters outside Hartford, Conn. (1955-57).

Lever Brothers turned to the same firm for New York City's Lever House (1952), in which the parklike plaza, glass-curtain walls, and thin aluminum mullions realized the dreams of Mies and others in the 1920s of freestanding crystalline shafts. Designed by Eliel Saarinen's son Eero, the General Motors Technical Center (1948-56) at Warren, Mich., was compared with Versailles in its extent, grandeur, and rigorous conformity to an austere, geometric aesthetic of Miesian forms.

The Harrison and Abramovitz's tower for the Aluminum Company of America at Pittsburgh (1954) advertised its own product, as did Skidmore, Owings, and Merrill's Inland Steel Building at Chicago (1955-57). Perhaps the chastest of all was the Seagram Building (1954-58) at New York City, designed by Mies and Philip Johnson. Wright alone avoided the rectilinear geometry of these office buildings. In 1955 he saw his Price Tower rise at Bartlesville, Okla., a richly faceted, concrete and copper fulfilment of the St. Mark's Tower he had designed more than 25 years earlier.

About 1952 there was a significant shift within Modernism from what had come to be called Functionalism, or the International Style, toward a monumental Formalism. There was increasing interest in highly sculptural masses and spaces, as well as in the decorative qualities of diverse building materials and exposed structural systems. Wright's Guggenheim Museum is a manifestation of this aesthetic.

Those who had focused their attention on the rectilinear portions of Le Corbusier's Savoye House and Unité d'Habitation apartments at Marseille (1946-52), tended to ignore the plastic sculpture on the roofs of those buildings; to such people, Le Corbusier's highly individual buildings at Chandigarh, India (begun 1950), and the cavernous space in the lyrical church of Notre-Dame-du-Haut at Ronchamp, Fr., seemed to be examples of personal whimsy.

Pier Luigi Nervi in Italy gave structural integrity to the complex curves and geometry of reinforced-concrete structures, such as the Orbetello aircraft hangar (begun 1938) and Turin's exposition hall (1948-50). The Spaniard Eduardo Torroja, his pupil Felix Candela, and the American Frederick Severud followed his lead.

Essentially, each attempted to create an umbrella roof the interior space of which could be subdivided as required, such as Torroja's grandstand for the Zarzuela racetrack in Madrid (1935). Mies constructed rectilinear versions of such a space in Crown Hall and in his Farnsworth House at Plano, Ill. (1946-50), while Philip Johnson allowed a single functional unit, the brick-cylinder utility stack, to protrude from his precise glass house at New Canaan, Conn. (1949).

Other designers used curvilinear structural geometry, best indicated by Matthew Nowicki's (1910-49) sports arena at Raleigh, N.C. (1952-53), in which two tilted parabolic arches, supported by columns, and a stretched-skin roof enclose a colossal space devoid of interior supports.

In 1949 Nowicki had challenged Louis Sullivan's precept, form follows function, with another, and form follows form, a dictum that freed architecture from programmatic expression. Hugh Stubbins' congress hall, at Berlin (1957), and Eero Saarinen's Trans World Airlines terminal at John F. Kennedy International Airport, New York City (1956-62), were outstanding examples of these dynamically monumental, single-form buildings the geometric shapes and silhouettes of which were derived from mathematical computation and technological innovation.

International competitions for the opera house at Sydney (1957) and a government centre at Toronto (1958) were won by the Dane Jørn Utzon and the Finn Viljo Revell, respectively. Both architects were exponents of the new monumentalism.

These designs posed problems in structural engineering and in scale, but many architects, such as the American Minoru Yamasaki in the McGregor Building for Wayne State University at Detroit (1958), attempted to make structure become decorative, while the decorative screen, as used by Edward Durell Stone at the U.S. embassy in New Delhi (1957-59), offered a device for wrapping programmatic interiors within a rich pattern of sculptured walls.

Mexico and South America broke their bonds to French, Spanish, and Portuguese academic design during the 1930s. Le Corbusier's influence became partially strong in Brazil, where the Brazilian Oscar Niemeyer and other architects designed the Corbusier-inspired Ministry of Education and Public Health at Rio de Janeiro (1937-42). Brazil's Lúcio Costa, Affonso Reidy, and Niemeyer; Mexico's Felix Candela, Juan O'Gorman, José Villagran Garcia, and Luis Barragán; and Venezuela's Carlos Raúl Villanueva were the vanguard of Latin-American architectural modernism.

Whole communities such as Caracas and São Paulo essentially were rebuilt during the 1950s, and new cities, such as Brasília, the capital of Brazil, and "university cities," such as those of Mexico and Venezuela, were conceived and erected. In Mexico there was avid support for modern design in buildings such as the Presidente Juárez housing at Mexico City (1950) by Mario Pani and Salvador Ortega.

In Colombia, after World War II, enormous strides were made in thin-shelled reinforced-concrete construction. In Brazil, dramatic complexes were erected from concrete by Reidy, such as the school and gymnasium at Pedregulho housing at Rio de Janeiro (1953) and Rio's Museum of Modern Art (1960-67).

After 1959, office buildings for administrative headquarters of large corporations followed the 1955-57 suburban-campus model of Skidmore, Owings, and Merrill's Connecticut General Life Insurance Company or, if urban, the towerlike form, often with strong structural expression (Torre Velasca, Milan, by Belgiojoso, Peressutti, and Rogers, 1959) or the slab form, usually emphasizing glazed walls (Mannesmann Building, Düsseldorf, by Paul Schneider-Esleben, 1959), but they rarely achieved an urban composition such as the 1962 Place Ville-Marie, built at Montreal by the Chinese-born American architect I.M. Pei.

Air transportation, trade exhibitions, and spectator sports summoned the often awesome spatial resources of modern technology. The stadiums for the 1964 Olympics at Tokyo by Tange Kenzo, Rome's Pallazzi dello Sport done by Nervi (1960), Eero Saarinen's Dulles International Airport at Chantilly, Va. (1958-62), and Chicago's exposition hall, McCormick Place, by C.F. Murphy and Associates (1971) are examples of the colossal spaces achieved in reinforced concrete or steel and glass.

International exhibitions seldom offered comparable architecture. At the New York World's Fair (1964) the Spanish pavilion by Javier Carvajal and the Japanese pavilion by Maekawa Kunio were buildings of merit. There were also several notable examples at Montreal's Expo 67: the West German pavilion by Frei Otto, the U.S. pavilion by R. Buckminster Fuller, and the startling Constructivist apartment house, Habitat 67, by the Israeli Moshe Safdie, in association with David, Barott, and Boulva, whose 158 precast-concrete apartment units were hoisted into place and post-tensioned to permit dramatic cantilevers and terraces.

World's fairs continued to provide a setting for occasionally distinguished examples of modern structures that demonstrated innovations in building technology.

The architecture of South and Southeast Asia as well as of Japan has been decisively influenced by Western architects, particularly Le Corbusier. The leading figure in Japan was Tange Kenzo, whose many powerful buildings of rough concrete include the Peace Centre, Hiroshima (1949-55), and St. Mary's Roman Catholic Cathedral at Tokyo (1965). His disciples included the so-called Metabolism Group, led by Kikutake Kiyonori, Maki Fumihiko, and Otaka Masato.

Their work, characterized by a dynamic science-fiction quality expressive of fluidity and change, culminated in the Osaka Expo 1970, with constructions such as Tange's giant space frame, known as the Theme Pavilion, and Kikutake's Landmark Tower.

Much significant architecture in the postwar period was sponsored by cultural centres and educational institutions, such as Berlin's philharmonic hall (1963) by Hans Scharoun. Louis I. Kahn, in his design for the Richards Medical Research Building (1960), gave the University of Pennsylvania in Philadelphia a linear programmatic composition of laboratories, each served by vertical systems for circulating gases, liquids, and electricity.

Paul Rudolph's art and architecture building (1963) at Yale University in New Haven, Conn., gathered its studios, galleries, classrooms, and

189

light wells on 36 interpenetrating levels distributed over six stories. The Morse and Stiles colleges (1962), also at Yale, were designed by Eero Saarinen and set a new standard for multiple-entry urban dormitories.

Even the traditionalist campuses of New England preparatory schools gained modern architecture, such as the art building and science building at Phillips Academy in Andover, Mass., by Benjamin A. Thomson (1963) and the dormitories at St. Paul's School in Concord, N.H., by Edward Larrabee Barnes (1965).

The innovations in educational architecture were international. In England, distinctive educational architecture arrived at Hunstanton Secondary School, Norfolk (1949-54), by Peter and Alison Smithson.

An example of what became known as the New Brutalism, this building was influenced by Mies van der Rohe. Most New Brutalist buildings, however, owed more to Le Corbusier's late work--for example, the gray concrete masses of Denys Lasdun's University of East Anglia, Norfolk (1962-68)--while James Stirling's History Faculty, Cambridge (1964-67), brought a neo-Constructivist element to the Brutalist tradition.

Canada gained the Central Technical School Arts Center by Robert Fairfield Associates (1964) and Scarborough College by John Andrews, with Page and Steele (1966), both at Toronto. Italian innovative educational architecture is exemplified in Milan's Instituto Marchiondi (1959) by Vittoriano Viganò. Led by disciples of Le Corbusier, the Japanese built Waseda University (1964), which was designed by Katsuo Ardo, and Maekawa Kunio's Gakushuin University (1964), both in Tokyo.

Some of the new educational settings proposed solutions to what was undoubtedly the mid-20th century's greatest problem, its urban environment. The high-rise, dense campus at Boston University by José Luis Sert and the skyscraper towers of MIT's earth-sciences building (1964) by I.M. Pei, as well as Harvard's behavioural sciences building (1964) by Minoru Yamasaki, were imaginative single buildings responding to urban circumstances.

The Air Force Academy at Colorado Springs, Colo., and the Chicago Circle Campus of the University of Illinois (1965), both by the firm of Skidmore, Owings, and Merrill with Walter A. Netsch as the principal designer (1956), and the Salk Institute for Biological Studies at La Jolla, Calif., by Louis I. Kahn (1966), all offered intimations of a new city built around a cultural, educational centre.

No comparable concentration of intensive, harmonious urban architecture was achieved for cities, even though, after 1955, the building of new cities produced some remarkable examples such as Vällingby, Swed.; Brasília, the new capital of Brazil; Cumbernauld, in Scotland; and Chandigarh, in India; and some remarkable renovations of old cities, as in Eastwicks in Philadelphia (Reynolds Metals Co.; plans by Constantinos Doxiadis, 1960) and Constitution Plaza in Hartford, Conn. (Charles DuBose, with Sasaki, Walker & Associates 1964), and New York's Lincoln Center for the Performing Arts (1962).

By this time, however, it was beginning to be felt that the application of Modern movement principles had caused visual damage to historic cities and had also failed to create a humane environment in new cities. It was at this moment that the postmodernist era began.

24.7 AERONAUTICAL ENGINEERING

The roots of aeronautical engineering can be traced to the early days of mechanical engineering, to inventors' concepts, and to the initial studies of aerodynamics, a branch of theoretical physics. The earliest sketches of flight vehicles were drawn by Leonardo da Vinci, who suggested two ideas for sustentation.

The first was an ornithopter, a flying machine using flapping wings to imitate the flight of birds. The second idea was an aerial screw, the predecessor of the helicopter. Manned flight was first achieved in 1783, in a hot-air balloon designed by the French brothers Joseph-Michel and Jacques-Étienne Montgolfier.

Aerodynamics became a factor in balloon flight when a propulsion system was considered for forward movement. Benjamin Franklin was one of the first to propose such an idea, which led to the development of the dirigible. The power-driven balloon was invented by Henri Gifford, a Frenchman, in 1852.

The invention of lighter-than-air vehicles occurred independently of the development of aircraft. The breakthrough in aircraft development came in 1799 when Sir George Cayley, an English baron, drew an airplane incorporating a fixed wing for lift, an empennage (consisting of horizontal and vertical tail surfaces for stability and control), and a separate propulsion system.

Because engine development was virtually nonexistent, Cayley turned to gliders, building the first successful one in 1849. Gliding flights established a data base for aerodynamics and aircraft design. Otto Lilienthal, a German scientist, recorded more than 2,000 glides in a

191

five-year period, beginning in 1891. Lilienthal's work was followed by the American aeronaut Octave Chanute, a friend of the American brothers Orville and Wilbur Wright, the fathers of modern manned flight.

Following the first sustained flight of a heavier-than-air vehicle in 1903, the Wright brothers refined their design, eventually selling airplanes to the U.S. Army. The first major impetus to aircraft development occurred during World War I, when aircraft were designed and constructed for specific military missions, including fighter attack, bombing, and reconnaissance.

The end of the war marked the decline of military high-technology aircraft and the rise of civil air transportation. Many advances in the civil sector were due to technologies gained in developing military and racing aircraft. A successful military design that found many civil applications was the U.S. Navy Curtiss NC-4 flying boat, powered by four 400-horsepower V-12 Liberty engines. It was the British, however, who paved the way in civil aviation in 1920 with a 12-passenger Handley-Page transport. Aviation boomed after Charles A. Lindbergh's solo flight across the Atlantic Ocean in 1927.

Advances in metallurgy led to improved strength-to-weight ratios and, coupled with a monocoque design, enabled aircraft to fly farther and faster. Hugo Junkers, a German, built the first all-metal monoplane in 1910, but the design was not accepted until 1933, when the Boeing 247-D entered service. The twin-engine design of the latter established the foundation of modern air transport.

The advent of the turbine-powered airplane dramatically changed the air transportation industry. Germany and Britain were concurrently developing the jet engine, but it was a German Heinkel He 178 that made the first jet flight on Aug. 27, 1939. Even though World War II accelerated the growth of the airplane, the jet aircraft was not introduced into service until 1944, when the British Gloster Meteor became operational, shortly followed by the German Me 262. The first practical American jet was the Lockheed F-80, which entered service in 1945.

Commercial aircraft after World War II continued to use the more economical propeller method of propulsion. The efficiency of the jet engine was increased, and in 1949 the British de Havilland Comet inaugurated commercial jet transport flight. The Comet, however, experienced structural failures that curtailed the service, and it was not until 1958 that the highly successful Boeing 707 jet transport began non-stop transatlantic flights.

While civil aircraft designs utilize most new technological advancements, the transport and general aviation configurations have changed only slightly since 1960. Because of escalating fuel and hardware prices, the development of civil aircraft has been dominated by the need for economical operation.

Technological improvements in propulsion, materials, avionics, and stability and controls have enabled aircraft to grow in size, carrying more cargo faster and over longer distances. While aircraft are becoming safer and more efficient, they are also now very complex. Today's commercial aircraft are among the most sophisticated engineering achievements of the day.

Smaller, more fuel-efficient airliners are being developed. The use of turbine engines in light general aviation and commuter aircraft is being explored, along with more efficient propulsion systems, such as the propfan concept.

Using satellite communication signals, onboard microcomputers can provide more accurate vehicle navigation and collision-avoidance systems. Digital electronics coupled with servo mechanisms can increase efficiency by providing active stability augmentation of control systems. New composite materials providing greater weight reduction; inexpensive one-man, lightweight, non-certified aircraft, referred to as ultra-lights; and alternate fuels such as ethanol, methanol, synthetic fuel from shale deposits and coal, and liquid hydrogen are all being explored.

Aircraft designed for vertical and short takeoff and landing, which can land on runways one-tenth the normal length, are being developed. Hybrid vehicles such as the Bell XV-15 tilt-rotor already combine the vertical and hover capabilities of the helicopter with the speed and efficiency of the airplane. Although environmental restrictions and high operating costs have limited the success of the supersonic civil transport, the appeal of reduced travelling time justifies the examination of a second generation of supersonic aircraft.

24.8 DISSEMINATION OF INFORMATION

The process of recording information by handwriting was obviously laborious and required the dedication of the likes of Egyptian scribes or monks in monasteries around the world. It was only after mechanical means of reproducing writing were invented that information records could be duplicated more efficiently and economically.

The first practical method of reproducing writing mechanically was block printing; it was developed in China during the T'ang dynasty (618-907). Ideographic text and illustrations were engraved in wooden blocks, inked, and copied on paper. Used to produce books as well as cards, charms, and calendars, block printing spread to Korea and Japan but apparently not to the Islamic or European Christian civilizations. European woodcuts and metal engravings date only to the 14th century.

Printing from movable type was also invented in China (in the mid-11th century AD). There and in the bookmaking industry of Korea, where the method was applied more extensively during the 15th century, the ideographic type was made initially of baked clay and wood and later of metal. The large number of typefaces required for pictographic text composition continued to handicap printing in the Orient until the present time.

The invention of character-oriented printing from movable type (1440-50) is attributed to the German printer Johannes Gutenberg. Within 30 years of his invention, the movable-type printing press was in use throughout Europe. Character-type pieces were metallic and apparently cast from metallic molds; paper and vellum (calfskin parchment) were used to carry the impressions.

Gutenberg's technique of assembling individual letters by hand was employed until 1886, when the German-born American printer Ottmar Mergenthaler developed the Linotype, a keyboard-driven device that cast lines of type automatically. Typesetting speed was further enhanced by the Monotype technique, in which a perforated paper ribbon, punched from a keyboard, was used to operate a type-casting machine.

Mechanical methods of typesetting prevailed until the 1960s. Since that time they have been largely supplanted by the electronic and optical printing techniques described in the previous section.

Unlike the use of movable type for printing text, early graphics were reproduced from wood relief engravings in which the nonprinting portions of the image were cut away. Musical scores, on the other hand, were reproduced from etched stone plates.

At the end of the 18th century the German printer Aloys Senefelder developed lithography, a planographic technique of transferring images from a specially prepared surface of stone. In offset lithography the image is transferred from zinc or aluminum plates instead of stone, and in photoengraving such plates are superimposed with film and then etched.

The first successful photographic process, the daguerreotype, was developed during the 1830s. The invention of photography, aside from providing a new medium for capturing still images and later video in analogue form, was significant for two other reasons. First, recorded information (textual and graphic) could be easily reproduced from film and, second, the image could be enlarged or reduced.

Document reproduction from film to film has been relatively unimportant, because both printing and photocopying (see above) are cheaper. The ability to reduce images, however, has led to the development of the microform, the most economical method of disseminating analogue-form information.

Another technique of considerable commercial importance for the duplication of paper-based information is photocopying, or dry photography. Printing is most economical when large numbers of copies are required, but photocopying provides a fast and efficient means of duplicating records in small quantities for personal or local use. Of the several technologies that are in use, the most popular process, xerography, is based on electrostatics.

While the volume of information issued in the form of printed matter continues unabated, the electronic publishing industry has begun to disseminate information in digital form. The digital optical disc (see above Recording media) is developing as an increasingly popular means of issuing large bodies of archival information--for example, legislation, court and hospital records, encyclopaedias and other reference works, referral databases, and libraries of computer software.

Full-text databases, each containing digital page images of the complete text of some 400 periodicals stored on CD-ROM, entered the market in 1990. The optical disc provides the mass production technology for publication in machine-readable form. It offers the prospect of having large libraries of information available in virtually every school and at many professional workstations.

The coupling of computers and digital telecommunications is also changing the modes of information dissemination. High-speed digital satellite communications facilitate electronic printing at remote sites; for example, the world's major newspapers and magazines transmit electronic page copies to different geographic locations for local printing and distribution. Updates of catalogues, computer software, and archival databases are distributed via electronic mail, a method of rapidly forwarding and storing bodies of digital information between remote computers.

Indeed, a large-scale transformation is taking place in modes of formal as well as informal communication. For more than three centuries, formal communication in the scientific community has relied on the scholarly and professional periodical, widely distributed to tens of thousands of libraries and to tens of millions of individual subscribers. In 1992 a major international publisher announced that its journals would gradually be available for computer storage in digital form; and in that same year the State University of New York at Buffalo began building a completely electronic, paperless library.

The scholarly article, rather than the journal, is likely to become the basic unit of formal communication in scientific disciplines; digital copies of such an article will be transmitted electronically to subscribers or, more likely, on demand to individuals and organizations who learn of its existence through referral databases and new types of alerting information services. The Internet already offers instantaneous public access to vast resources of non-commercial information stored in computers around the world.

Similarly, the traditional modes of informal communications--various types of face-to-face encounters such as meetings, conferences, seminars, workshops, and classroom lectures--are being supplemented and in some cases replaced by electronic mail, electronic bulletin boards (a technique of broadcasting newsworthy textual and multimedia messages between computer users), and electronic teleconferencing and distributed problem-solving (a method of linking remote persons in real time by voice-and-image communication and special software called "groupware").

These technologies are forging virtual societal networks--communities of geographically dispersed individuals who have common professional or social interests.

24.9 HELICOPTER - HISTORY

One important characteristic of the history of vertical flight is the pervasive human interest in the subject; inventors in many countries took up the challenge over the years, achieving varying degrees of success. The history of vertical flight began at least as early as about AD 400; there are historical references to a Chinese kite that used a rotary wing as a source of lift.

Toys using the principle of the helicopter--a rotary blade turned by the pull of a string--were known during the Middle Ages. During the latter part of the 15th century, Leonardo da Vinci made drawings of a

196

helicopter that used a spiral airscrew to obtain lift. A toy helicopter, using rotors made out of the feathers of birds, was presented to the French Academy of Science in 1784 by two artisans, Launoy and Bienvenu; this toy forecast a more successful model created in 1870 by Alphonse Pénaud in France.

The first scientific exposition of the principles that ultimately led to the successful helicopter came in 1843 from Sir George Cayley, who is also regarded by many as the father of fixed-wing flight. From that point on, a veritable gene pool of helicopter ideas was spawned by numerous inventors, almost entirely in model or sketch form. Many were technical dead ends, but others contributed a portion of the ultimate solution. In 1907 there were two significant steps forward.

On September 29, the Breguet brothers, Louis and Jacques, under the guidance of the physiologist and aviation pioneer Charles Richet made a short flight in their Gyroplane No. 1, powered by a 45-horsepower engine. The Gyroplane had a spiderweb-like frame and four sets of rotors.

The piloted aircraft lifted from the ground to a height of about two feet, but it was tethered and not under any control. Breguet went on to become a famous name in French aviation, and in time Louis returned to successful work in helicopters.

Later, in November, their countryman Paul Cornu, who was a bicycle maker like the Wright brothers, attained a free flight of about 20 seconds' duration, reaching a height of one foot in a twin-rotor craft powered by a 24-horsepower engine. Another man who, like the Breguets, would flirt with the helicopter, go on to make his name with fixed-wing aircraft, and then later return to the challenge of vertical flight, was Igor Sikorsky, who made some unsuccessful experiments at about the same time.

The next 25 years were characterized by two main trends in vertical flight. One was the wide spread of minor successes with helicopters; the second was the appearance and apparent success of the autogiro (also spelled autogyro).

The helicopter saw incremental success in many countries, and the following short review will highlight only those whose contributions were ultimately found in successfully developed helicopters. In 1912 the Danish inventor Jacob Ellehammer made short hops in a helicopter that featured contrarotating rotors and cyclic pitch control, the latter an important insight into the problem of control.

197

On Dec. 18, 1922, a complex helicopter designed by George de Bothezat for the U.S. Army Air Force lifted off the ground for slightly less than two minutes, under minimum control. In France on May 4, 1924, Étienne Oehmichen established a distance record for helicopters by flying a circle of a kilometre's length.

In Spain in the previous year, on Jan. 9, 1923, Juan de la Cierva made the first successful flight of an autogiro. An autogiro operates on a different principle than a helicopter. Its rotor is not powered but obtains lift by its mechanical rotation as the autogiro moves forward through the air. It has the advantage of a relatively short takeoff and a near vertical descent, and the subsequent success of Cierva's autogiros and those of his competitors seemed to cast a pall on the future of helicopter development.

Autogiros were rapidly improved and were manufactured in several countries, seeming to fill such a useful niche that they temporarily overshadowed the helicopter. Ironically, however, the technology of the rotor head and rotor blade developed for the autogiro contributed importantly to the development of the successful helicopter, which in time made the autogiro obsolete.

In 1936 Germany stepped to the forefront of helicopter development with the Focke Achgelis Fa 61, which had two three-bladed rotors mounted on outriggers and powered by a 160-horsepower radial engine. The Fa 61 had controllable cyclic pitch and set numerous records, including, in 1938, an altitude flight of 11,243 feet and a cross-country flight of 143 miles.

In 1938 the German aviator Hanna Reitsch became the world's first female helicopter pilot by flying the Fa 61 inside the Deutschland-Halle in Berlin. It was both a technical and a propaganda triumph. Germany continued its helicopter development during World War II and was the first to place a helicopter, the Flettner Kolibri, into mass production.

In the United States, after many successes with commercial flying boats, Igor Sikorsky turned his attention to helicopters once again, and after a long period of development he made a successful series of test flights of his VS-300 in 1939-41. Essentially a test aircraft designed for easy and rapid modification, the VS-300 was small (weighing 1,092 pounds) and was powered by a 65-horsepower Lycoming engine.

Yet it possessed the features that characterize most modern helicopters: a single main three-bladed rotor, with collective pitch, and a tail rotor. As successful as the VS-300 was, however, it also clearly

198

showed the difficulties that all subsequent helicopters would experience in the development process.

For many years, compared with conventional aircraft, helicopters were underpowered, difficult to control, and subject to much higher dynamic stresses that caused material and equipment failures. Yet the VS-300 led to a long line of Sikorsky helicopters, and it influenced their development in a number of countries, including France, England, Germany, and Japan.

After World War II the commercial use of helicopters developed rapidly in many roles, including fire fighting, police work, agricultural crop spraying, mosquito control, medical evacuation, and carrying mail and passengers.

The expanding market brought additional competitors into the field, each with different approaches to the problem of vertical flight. The Bell Aircraft Corporation, under the leadership of Arthur Young, began its long, distinguished history of vertical-flight aircraft with a series of prototypes that led to the Bell Model 47, one of the most significant helicopters of all time, incorporating an articulated, gyro-stabilized, two-blade rotor.

Frank Piasecki created the Piasecki Helicopter Corporation; its designs featured a tandem-rotor concept. The use of twin tandem rotors enabled helicopters to grow to almost twice their previous size without the difficulty of creating very large rotor blades. In addition, the placement of the twin rotors provided a large centre of gravity range. The competition was international, with rapid progress made in the Soviet Union, the United Kingdom, France, Italy, and elsewhere.

To an even greater extent than fixed-wing aircraft, the development of the helicopter had been limited by engine power. Reciprocating engines were heavy, noisy, and less efficient at high altitude. The first application of jet-engine technology to the helicopter was accomplished in 1951 by the Kaman Aircraft Corporation's HTK-1, which had Kaman's patented aerodynamic servo-controlled rotors in the "synchropter" configuration (*i.e.,* side-by-side rotors with intermeshing paths of blade travel).

In conventional aircraft the power of the jet engine was used primarily for increased speed. In the helicopter the thrust of the jet turbine had to be captured by a gearbox that would turn the rotor. The jet engine had many advantages for the helicopter--it was smaller, weighed less than a piston engine of comparable power, had far less vibration, and used less expensive fuel. The French SNCA-S.E. 3130 Alouette II made its first flight on March 12, 1955, powered by a Turbomeca Artouste II turbine engine. It rapidly became one of the most influential helicopters in the world and started a trend toward jet-powered helicopters everywhere.

There are now a vast number of helicopter types available on the market, ranging from small two-person private helicopters through large passenger-carrying types to work vehicles capable of carrying huge loads to remote places. All of them respond to the basic principles of flight, but, because of the unique nature of the helicopter's rotor and control systems, the techniques for flying them differ.

There are other types of vertical-lift aircraft, whose controls and techniques are often a blend of the conventional aircraft and the helicopter. They form a small part of the total picture of flight but are of growing importance.

24.10 OPTOELECTRONICS

Many semiconductor materials other than silicon and germanium exist, and they have different useful properties. Compounds formed by the elements from column III of the periodic table--such as aluminum, gallium, and indium--with those from column V--such as phosphorus, arsenic, and antimony--are of particular interest. These so-called III-V compounds are used to make semiconductor devices that emit light efficiently or that operate at exceptionally high frequencies.

A remarkable characteristic of these compounds is that they can, in effect, be mixed together. One can produce gallium arsenide, or substitute aluminum for some of the gallium, or also substitute phosphorus for some of the arsenic. When this is done, the electrical and optical properties of the material are subtly changed in a continuous fashion in proportion to the amount of aluminum or phosphorus used.

All these compounds have the same crystal structure. This makes possible the gradation of composition, and thus the properties, of the semiconductor material within one continuous crystalline body. Modern material-processing techniques allow these compositional changes to be controlled accurately on an atomic scale.

These characteristics are exploited in making semiconductor lasers that produce light of any given wavelength within a considerable range. Such lasers are used, for example, in compact digital audio disc players and as light sources for optical fibre communication.

A new direction in electronics employs photons (packets of light) instead of electrons. By common consent these new approaches are included in electronics, because the functions that are performed are, at least for the present, the same as those performed by electronic systems and because these functions usually are embedded in a largely electronic environment. This new direction is called optical electronics, or optoelectronics.

In 1966 it was proposed on theoretical grounds that glass fibres could be made with such high purity that light could travel through them for great distances. Such fibres were produced during the early 1970s. They contain a central core in which the light travels.

The outer cladding is made of glass of a different chemical formulation and has a lower optical index of refraction. This difference in refractive index indicates that light travels faster in the cladding than it does in the core. Thus, if the light beam begins to move from the core into the cladding, its path is bent so as to move it back into the core. The light is constrained within the core even if the fibre is bent into a circle.

The core of early optical fibres was of such a diameter (several micro-metres, or about one-tenth the diameter of a human hair) that the various rays of light in the core could travel in slightly different paths, the shortest directly down the axis and other longer paths wandering back and forth across the core. This limited the maximum distance that a pulse of light could travel without becoming unduly spread by the time it arrived at the receiving end of the fibre, with the central ray arriving first and others later.

In a digital communications system, successive pulses can overlap one another and be indistinguishable at the receiving end. Such fibres are called multimode fibres, in reference to the various paths (or modes) that the light can follow.

During the late 1970s, fibres were made with smaller core diameters in which the light was constrained to follow only one path. This occurs if the core has a diameter about the same as the wavelength of the light travelling in it--*i.e.*, about 1 to 2 micro-metres (0.001 to 0.002 millimetre, or 0.000039 to 0.000078 inch).

These single-mode fibres avoid the difficulty described above. By 1993 optical fibres capable of carrying light signals more than 215 kilometres (135 miles) became available. Specialized fibres that incorporate integral amplifying features show promise of being able to carry light signals over transoceanic distances without the need for conventional pulse regeneration measures.

201

Optical fibres have several advantages over the copper wires or coaxial cables so widely used in the past. They can carry information at a much higher rate, they occupy less space (an important feature in large cities and in buildings), and they are quite insensitive to electrical noise. Moreover, it is virtually impossible to make unauthorized connections to them.

Costs, initially high, had dropped by 1985 to the point where most new installations of telephone circuits between central telephone offices and longer distances consisted of optical fibres.

Most current installations use a single light signal travelling in one direction within an optical fibre. The light is provided by a solid-state laser and detected at the receiving end by a semiconductor diode. There is no reason that more than one light signal cannot be present at one time in a fibre; many such signals have been sent down a single fibre in laboratory tests.

Each signal is of a slightly different wavelength and can be separated from the others at the receiving end. Signals also have been sent in both directions simultaneously in the laboratory. New terminal equipment can be retrofitted to allow fibres now in service to carry much more information than they were originally intended to do, and this is in fact beginning to occur. The cost of long-distance communication can be significantly reduced, thereby encouraging the use of these circuits for more purposes than at present.

A second phase of optoelectronics was being developed during the late 1980s, but the improved system was not expected to be in service for several years. Given the fact that communication signals arrive at a central switching office in optical form, it is attractive to consider switching them from one route to another by optical means rather than electrically, as is done today.

The distances between central offices in most cases are substantially less than the distance light can travel within a fibre. Optical switching would make unnecessary the detection and regeneration of the light signals, steps that are currently required. The principles of an optical central-office switch are already understood, though much research is still needed to provide the new optical components and new manufacturing technology required to produce such a switch.

A third direction in optoelectronics builds in part on the foregoing developments but to a quite different end. A key problem in developing faster computers and faster integrated circuits to use in them is related to the time required for electrical signals to travel over wire interconnections. This is a difficulty both for the integrated circuits themselves as well as for the connections between them. Under the best circumstances, electrical signals can travel in a wire at about 90 percent of the speed of light.

A more usual rate is 50 percent. Light travels about 30 centimetres in a billionth of a second. Modern supercomputers operate at speeds of more than 1 billion operations per second. Thus, if two signals that start simultaneously from different sites are to arrive at their destination simultaneously, the paths they travel must not differ in length by more than a few centimetres.

Two approaches can be envisioned. In one, all the integrated circuits are placed as close together as possible to minimize the distances that signals must travel. This creates a cooling problem, because the integrated circuits generate heat. In the other possible approach, all the paths for signals are made equal to the longest path. This requires the use of much more wire, because most paths are longer than they would otherwise be. All this wire takes space, which means that the integrated circuits have to be placed farther apart than is preferable.

Ultimately, as computers operate even faster, neither approach will work, nor a radically new technique must be used. Optical communication between integrated circuits is one possible answer. Light beams do not take up space or interfere with cooling air. If the communication is optical, then the computation might be done optically as well.

Optical computation will require a radically different form of integrated circuit. Such integrated circuits can in principle be made of gallium arsenide and related III-V compounds. Some of these integrated circuits may be useful in an optical central-office switch. These matters are currently under serious study in research laboratories.

24.11 MOTION PICTURE, HISTORY OF

Thomas Alva Edison invented the phonograph in 1877 and it had quickly become the most popular home entertainment device of the century. It was to provide a visual accompaniment to the phonograph that Edison commissioned Dickson, a young laboratory assistant, to invent a motion-picture camera in 1887.

Dickson built upon the work of Muybridge and Marey, a fact that he readily acknowledged, but he was the first to combine the two final essentials of motion-picture camera and projection technology. These were a device, adapted from the escapement mechanism of a clock, to

203

ensure the intermittent but regular motion of the film strip through the camera and a regularly perforated celluloid film strip to ensure precise synchronization between the film strip and the shutter.

Dickson's camera was patented as the Kinetograph in 1893, and it initially imprinted up to 50 feet of celluloid film at the rate of about 40 frames per second.

Dickson was not the only person who had been tackling the problem of recording and reproducing moving images. Inventors throughout the world had been trying for years to devise working motion-picture machines. In fact, several European inventors, including the French-born Louis Le Prince and the Englishman William Friese-Greene, applied for patents on various cameras, projectors, and camera-projector combinations contemporaneously or even before Edison and his associates did. These machines were unsuccessful for a number of reasons, however, and little evidence survives of their actual practicality or workability.

Because Edison had originally conceived of motion pictures as an adjunct to his phonograph, he did not commission the invention of a projector to accompany the Kinetograph. Rather, he had Dickson design a type of peep-show viewing device called the Kinetoscope in which a continuous 47-foot film loop ran on spools between an incandescent lamp and a shutter for individual viewing.

Starting in 1894, Kinetoscopes were marketed commercially through the firm of Raff and Gammon for $250 to $300 apiece, and the Edison Company established its own Kinetograph studio (a single-room building called the "Black Maria" that rotated on tracks to follow the sun) in West Orange, N.J., to supply films for the Kinetoscopes that Raff and Gammon were installing in penny arcades, hotel lobbies, amusement parks, and other such semi-public places. In April of that year the first Kinetoscope parlour was opened in a converted storefront in New York City. The parlour charged 25 cents for admission to a bank of five machines.

The syndicate of Maguire and Baucus acquired the foreign rights to the Kinetoscope in 1894 and began to market the machines. Edison had declined to file for international patents on either his camera or his viewing device, and as a result the machines were widely and legally copied throughout Europe, where they were modified and improved far beyond the American originals.

In fact, it was a Kinetoscope exhibition in Paris that inspired the Lumière brothers, Auguste and Louis, to invent the first commercially viable projector. Their *cinématographe*, which functioned as a camera

204

and printer as well as a projector, ran at the economical speed of 16 frames per second. It was given its first commercial demonstration on Dec. 28, 1895.

Unlike the Kinetograph, which was battery-driven and weighed more than 1,000 pounds, the *cinématographe* was hand-cranked, lightweight (less than 20 pounds), and relatively portable. This naturally affected the kinds of films that were made with each machine: Edison films initially featured material such as circus or vaudeville acts that could be brought into a small studio and played out before an inert camera, while early Lumière films were mainly documentary views, or "actualities," shot outdoors on location.

In both cases, however, the films themselves were composed of a single, unedited shot emphasizing lifelike movement; they contained little or no narrative content. (After a few years design changes in the machines made it possible for Edison and the Lumières to shoot the same kinds of subjects.)

In general, Lumière technology became the European standard during the early primitive era, and because the Lumières sent their cameramen all over the world in search of exotic subjects, the *cinématographe* became the founding instrument of such far-flung cinemas as the Russian, the Australian, and the Japanese. (It also, of course, is the source for the word cinema.)

In the United States, the Kinetoscope installation business had reached saturation point by the summer of 1895, although it was still quite profitable for Edison as a supplier of films. Raff and Gammon persuaded Edison to buy the rights to a state-of-the-art projector, developed by Thomas Armat of Washington, D.C., which incorporated a superior intermittent movement mechanism and a loop-forming device (known as the Latham loop, after its earliest promoters, Grey and Otway Latham) to reduce film breakage, and in early 1896 Edison began to manufacture and market this machine as his own invention.

Given its first public demonstration on April 23, 1896, at Koster and Bial's Music Hall in New York City, the Edison Vitascope brought projection to the United States and established the format for American film exhibition for the next several years. It also encouraged the activities of such successful Edison rivals as the American Mutoscope and Biograph Company, which was formed in 1896 to exploit the Mutoscope peep-show device and the American Biograph camera and projector patented by W.K.L. Dickson in 1896.

During this time, which has been characterized as the "novelty period," emphasis fell on the projection device itself, and films

205

achieved their main popularity as self-contained vaudeville attractions. Vaudeville houses, among which there was intense competition at the turn of the century, headlined the name of the machines rather than the films (The Vitascope--Edison's Latest Marvel, The Amazing Cinématographe).

The projectors came supplied from the producer, or manufacturer, with an operator and a program of shorts. These films, whether they were Edison-style theatrical variety shorts or Lumière-style actualities, were perceived by their original audiences not as motion pictures in the modern sense of the term but as "animated photographs" or "living pictures," emphasizing their continuity with more familiar media of the time.

During the novelty period, the film industry was autonomous and unitary, with production companies leasing a complete film service of projector, operator, and shorts to the vaudeville market as a single, self-contained act. Starting around 1897, however, manufacturers began to sell both projectors and films to itinerant exhibitors who travelled with their programs from one temporary location (vaudeville theatres, fairgrounds, circus tents, lyceums) to another as the novelty of their films wore off at a given site.

This new mode of screening by circuit marked the first separation of exhibition from production and gave the exhibitors a large measure of control over early film form, since they were responsible for arranging the one-shot films purchased from the producers into audience-pleasing programs.

The putting together of these programs--which often involved narration, sound effects, and music--was in effect a primitive form of editing, so that it is possible to regard the itinerant projectionists working between 1896 and 1904 as the earliest directors of motion pictures. Several of them, notably Edwin S. Porter, were, in fact, hired as directors by production companies after the industry had stabilized in the first decade of the 20th century.

By encouraging the practice of peripatetic exhibition, the U.S. producers' policy of outright sales inhibited the development of permanent film theatres in the United States until nearly a decade after their appearance in Europe, where England and France had taken an early lead in both production and exhibition. Britain's first projector, the theatrograph (later, the animatograph), had been demonstrated in 1896 by the scientific instrument maker Robert W. Paul.

In 1899 Paul formed his own production company for the manufacture of actualities and trick films, and until 1905 Paul's Animatograph Works, Ltd., was England's largest producer, turning out an average of 50 films per year. Between 1896 and 1898, two Brighton photographers, George Albert Smith and James Williamson, constructed their own motion-picture cameras and began producing trick films featuring superimpositions (*The Corsican Brothers*, 1897) and interpolated close-ups (*Grandma's Reading Glass*, 1900; *The Big Swallow*, 1901).

Smith subsequently developed the first commercially successful photographic colour process (Kinemacolor, *c*. 1906-08, with Charles Urban), while Williamson experimented with parallel editing as early as 1900 (*Attack on a Chinese Mission Station*) and became a pioneer of the chase film (*Stop Thief !*, 1901; *Fire!*, 1901).

Both Smith and Williamson had built studios at Brighton by 1902 and, with their associates, came to be known as members of the "Brighton school," although they did not represent a coherent movement. Another important early British filmmaker was Cecil Hepworth, whose *Rescued by Rover* (1905) is regarded by many historians as the most skillfully edited narrative produced before the Biograph shorts of D.W. Griffith.

24.12 CRYPTOLOGY

During the first two years of World War I, the belligerents employed cipher systems almost exclusively for tactical communications; code systems were still used mainly for high-command and diplomatic communications. Field cipher systems, however, such as the U.S. Signal Corps cipher disk, lacked sophistication. Nevertheless, some complicated cipher systems were used for high-level communications by the end of the war, the most famous of which was the German ADFGVX fractionation cipher.

The communications needs of telegraphy and radio and the maturing of mechanical and electromechanical technology came together in the 1920s to bring about a true revolution in crypto-devices: the development of rotor cipher machines.

Although the concept of the rotor had been anticipated in the older mechanical cipher disks, the credit goes to an American, Edward H. Hebern, for first recognizing that by hardwiring a mono=alphabetic substitution in the connections from the contacts on one side of an electrical rotor to those on the other side and cascading a collection of

207

such rotors, poly-alphabetic substitutions of almost arbitrary complexity could be realized.

Hebern also recognized that a permutation in which several letters were shifted by the same amount in the rotor connections, say A to D and B to E, was cryptographically weaker than one in which this partial periodicity was minimized and designed his rotors accordingly.

Starting in 1921 and continuing through the next decade, Hebern constructed a series of steadily improving rotor machines that were evaluated by the U.S. Navy and undoubtedly led to the United States' superior position in cryptology as compared to the Axis powers during World War II.

The 1920s were marked by a series of challenges by inventors of cipher machines to national cryptologic services and by one service to another, resulting in a steady improvement both of cryptomachines and of cryptanalytic techniques for the analysis of machine ciphers.

At almost the same time that Hebern was inventing the rotor cipher machine in the United States, European engineers, notably Hugo A. Koch of The Netherlands and Arthur Scherbius of Germany, independently discovered the rotor concept and designed machines that became the precursors of the best known cipher machine in history, the German Enigma used in World War II.

By an indirect path of development, these machines were the stimulus for the TYPEX, the cipher machine employed by the British during World War II. The United States introduced the M-134-C (SIGABA) cipher machine during World War II.

The Japanese cipher machines of World War II have an interesting history linking them to both the Hebern machines and the Enigma. The Washington Conference on naval disarmament (1921-22) had as a primary objective limiting the total tonnage of capital ships (battleships, cruisers, and aircraft carriers) by the major powers--the United States, Great Britain, Japan, France, and Italy.

The most difficult problem was the way in which this tonnage was to be allocated among the five countries. The Japanese Foreign Office sent detailed cipher instructions to its ambassador in Washington, D.C., to negotiate for a 10-to-7 U.S.-to-Japanese tonnage ratio, to fall back to 10 to 6.5 if that failed, and only as a last resort to retreat to a lowest acceptable ratio of 10 to 6. In a flash of inspiration Herbert O. Yardley broke the Japanese ciphers (see above Basic aspects), enabling the U.S. representative, Secretary of State Charles Evans Hughes, to press for this lower limit.

The Japanese reluctantly accepted the inferior position of 10:10:6:3.3:3.3 (the United States, Britain, Japan, France, and Italy, respectively) laid out in the Five-Powers Treaty. Primarily because of a failure of their cryptography, they had settled for 100,000 tons of shipping less than they might otherwise have obtained, a difference of three capital ships. When Yardley later revealed in 1928 and subsequently published in *The American Black Chamber* the details of the American successes in cryptanalyzing the Japanese ciphers, with the associated costs to Japan, the Japanese government set out to develop the best cryptomachines possible.

With this end in mind, it purchased the rotor machines of Hebern and Hagelin and the commercial Enigmas, as well as several other contemporary machines, for study and analysis. In 1930 the Japanese Foreign Office put into service its first rotor machine, which was code-named RED by U.S. cryptanalysts. In 1935-36 the U.S. Army Signal Intelligence Service (SIS) team of cryptanalysts, led by William F. Friedman, succeeded in cryptanalyzing RED ciphers, drawing heavily on its previous experience in cryptanalyzing the machine ciphers produced by the Hebern rotor machines.

It was an ironic twist of fate that the Hebern machines, which were never commercially successful, played such a pivotal role in the design of two widely used rotor machines and in the evolution of the techniques that were vital to the cryptanalysis of the RED ciphers. In 1939 the Japanese introduced a new cipher machine, code-named PURPLE by U.S. cryptanalysts, in which rotors were replaced by telephone stepping switches.

Because the replacement of RED machines by PURPLE machines was gradual, providing an enormous number of cribs between the systems to aid cryptanalysts, and because the Japanese had taken a shortcut to avoid the key distribution problem by generating keys systematically, U.S. cryptanalysts were able not only to cryptanalyze the RED ciphers but also to anticipate keys several days in advance.

Functionally equivalent PURPLE cipher machines were constructed by Friedman and his SIS associates and used throughout the war to decrypt Japanese ciphers. Apparently no PURPLE machine survived the war. Another Japanese cipher machine code-named JADE was essentially the same as the PURPLE. It differed from the latter chiefly in that it typed Japanese kana characters directly.

The greatest triumphs in the history of cryptanalysis were the Polish and British solution of the German Enigma ciphers and of two teleprinter ciphers, code-named ULTRA, and the American cryptanalysis of the Japanese RED, ORANGE, and PURPLE ciphers, code-named MAGIC. These developments played a major role in the Allies' conduct of World War II. Of the two, the cryptanalysis of the Japanese ciphers is the more impressive technically, because it was a tour de force of cryptanalysis against ciphertext alone.

In the case of the Enigma machines, the basic patents had been issued in the United States, commercial machines were widely available, and the rotor designs were known to Allied cryptanalysts from a German code clerk. Although such factors do not diminish the practical importance of the ULTRA intercepts, they did make the cryptanalysis easier

24.13 AGE OF THE AIRCRAFT CARRIER

Early in World War II the primary instrument for delivering naval combat power became the aircraft carrier. The reason was range: aircraft could deliver a concerted attack at 200 miles or more, whereas battleships could do so only at 20 miles or less. The foremost tactical question during the transition in the 1920s and '30s was whether aircraft could lift enough destruction to supersede the battleship. Into the 1930s sceptics were correct that aircraft could not.

But by the end of that decade, engines were carrying adequate payloads, dive-bomber and torpedo-plane designs had matured, carrier arresting gear and associated flight-deck handling facilities were up to their tasks, and proficient strike tactics had been well practiced. U.S. and Japanese naval aviators were pacesetters in these developments.

There was a subordinate tactical question as well: could the enemy be found at the outer limits of aircraft range? The ability to attack fixed targets such as the Panama Canal or Pearl Harbor, and to achieve surprise in doing so, had been amply demonstrated in naval exercises as well as in battle, but finding, reporting, and closing on ships at sea was a greater challenge.

Without detracting from the courage and skill of aviators, it may be said that effective scouting was the dominant tactical problem of carrier warfare and had utmost influence on the outcomes of the crucial carrier battles of the Pacific Theatre in 1942: the Coral Sea (May 4-8), Midway (June 3-6), the Eastern Solomons (August 23-25), and the Santa Cruz Islands (October 26). In those closely matched battles the quality of U.S. and Japanese aviators and their planes was virtually on a par. When the United States won, it did so by superior scouting and screening, owing in large measure to air-search radar and to the advantage of having broken the Japanese code.

The command and control structure polished by the U.S. Navy during the war was the third vital component, after scouting and the delivery of firepower. The tangible manifestation of modern C^2 was the Combat Information Centre, which centralized radar information and voice radio communications.

By 1944 the tactical doctrine of coordinating fighter air defences, along with the now much strengthened antiaircraft firepower on ships of the fleet, was so effective that in the Battle of the Philippine Sea (June 19-21, 1944) more than 90 percent of 450 Japanese aircraft were wiped out in a fruitless attack on Admiral Raymond Spruance's 5th Fleet.

The new tactical formation was circular, with carriers in the centre defended by an antiaircraft and antisubmarine screen composed of their own aircraft plus battleships, cruisers, and destroyers. For offensive purposes, a circle allowed a rapid simultaneous turn by all ships in a task group in order to launch and recover aircraft. For antiaircraft defence, the circle was shrunk in diameter as tightly as possible so that each screening ship, by defending itself, helped defend its neighbour.

The new battle paradigm called for a pulse of combat power to be delivered in a shock attack by one or more air wings. Despite every intention, though, air strikes against alerted defences were rarely delivered as compactly as practiced, nor were they as decisive tactically as naval aviators had expected. In the five big carrier battles, one attacking air wing took out an average of only one enemy carrier. (Viewed strategically, this average, along with losses of aircraft of around 50 percent per battle, was enough to govern the pattern of the Pacific war.)

Since it took more than two hours to launch, marshal, and deliver an air strike, it was difficult to attack before an enemy counter-strike was in the air. Successful command at sea depended as never before on effective scouting and communication, because in order to win a decisive battle, in World War II as in all of naval history, it was necessary to attack effectively first.

Dominant though it was, carrier-based air power did not control the seas at night. With a modicum of success, the high-quality ships of Germany exploited the hours of darkness, especially during the winter months and in northern waters. In the bitterly contested campaign for Guadalcanal in the fall of 1942, guns ruled supreme at night and very nearly tipped the balance in favour of Japan. Expecting to be outnumbered as a result of the Five-Power Naval Limitation Treaty of 1922, the Imperial Japanese Navy had practiced night tactics assiduously in order, as they supposed, to whittle down the U.S. battle line during its slow march west across the Pacific.

Having developed the matchless Long Lance torpedo, they installed it liberally in light cruisers and destroyers and developed tactics that would hurl a barrage of the long-range weapons in the direction of the enemy line--at the same time taking care not to expose the beams of their own ships to a counterstroke. Standard U.S. doctrine, on the other hand, called for fighting in column, employing guns as the primary weapon; the advantages that should have accrued to the Americans at night from superior radar were largely squandered.

Between August 1942 and July 1943, in the cruiser-destroyer battles of Savo Island, Cape Esperance, Tassafaronga, Kula Gulf, and Kolombangara, Japanese night tactics prevailed. Not until mid-1943, with tactics attributed to Captain (later Admiral) Arleigh Burke that exploited the radar advantage in full, did the U.S. Navy redress the balance.

Still, naval aircraft were the weapons of decision. Although the duels of the great carrier fleets received more attention, air strikes from sea to shore were as crucial in securing control of the seas. Strikes by the British at Taranto, Italy (Nov. 11, 1940), by the Japanese at Pearl Harbor (Dec. 7, 1941), and by the Americans in the South Pacific at Rabaul (Nov. 5 and 11, 1943) and Truk (Feb. 17-18, 1944) were as important to that end as were the more sensational fleet engagements.

Also, in 1944 and 1945 the U.S. 3rd and 5th fleets, 27 fast carriers strong, took the war successfully against entire complexes of airfields in Formosa (now Taiwan), the Philippines, and Japan itself. A traditional tactical maxim, "Ships do not fight forts," was suspended for the duration of the war.

In the closing days of the war in the Pacific, the Battle of Okinawa served to indicate the nature of future combat at sea. By that time the U.S. Navy had reduced the Japanese Navy to impotence, and manned aircraft could not penetrate the sure American defences.

Nevertheless, during the three-month campaign for Okinawa (April-June 1945) the U.S. Navy lost 26 ships and suffered damage to 164 more--this time to Japanese kamikazes (suicide pilots) flying out of airfields in Japan. The pilots who flew these one-way missions were delivering, in effect, human guided missiles.

Kamikazes showed that missiles could, on sufficient occasion, get through otherwise impenetrable defences. The missile-guidance technology exhibited in the late stages of the war in Europe indicated that missiles would be the kamikazes of the future. And the atomic bomb offered the ugly threat of "one hit, one kill" at sea.

24.14 TYPOGRAPHY

Two late-19th-century developments--one technological, the other aesthetic--profoundly changed the course of book typography and design. The advent of mechanical type composition in the 1880s (the so-called Linotype machine was patented by Ottmar Mergenthaler, a German inventor, in 1884; the Monotype, by an American, Tolbert Lanston, in 1887) had much to do with the look of the 20th century book.

The Arts and Crafts Movement, whose leader in typography as in other aspects was William Morris, had an equally great influence on the quality of modern book printing.

24.15 PRIVATE-PRESS MOVEMENT

The Industrial Revolution changed the course of printing not only by mechanizing a handicraft but also by greatly increasing the market for its wares. Inventors in the 19th century, in order to produce enough reading matter for a constantly growing and ever more literate population, had to solve a series of problems in paper production, composition, printing, and binding.

The solution that most affected the appearance of the book was mechanical composition: the new composing machines imposed new limitations not only on type design but also on the number and kinds of faces available, since the money required to buy a new typeface was enough to inhibit printers from stocking faces of slight utility. As a result, Victorian exuberance of design, which might use a dozen or more typefaces within a single book, was effectively curbed.

It is paradoxical that what became known as the Arts and Crafts Movement, with its roots in the romantic Gothicism propounded by the critic John Ruskin and by Morris, should have had a considerable influence on modern industrial design, including that of the book. An Englishman, William Morris was a fervent Socialist who believed that the Industrial Revolution had killed man's joy in his work and that mechanization, by destroying handicraft, had brought ugliness with it.

213

Morris was above all a decorator; his work in the decorative arts had added great lustre to the fame he had already achieved as a writer when, partly as a result of dissatisfaction with the editions of his own works, he decided to establish a press. In 1888 Morris attended a lecture given by the printer Emery (later Sir Emery) Walker and was entranced by Walker's lantern slides of early types, greatly enlarged. He proposed to Walker that they cut a new font of type that would recapture the strength and beauty of the early letters, based upon medieval calligraphy.

The Kelmscott Press, in its brief life (1891-96), printed 52 books that exemplified Morris' standards of perfect workmanship. A firm believer that a return to the past would produce a better society, he commissioned handmade paper like that used in the 15th century, had new, blacker inks made, and used the hand-press and hand binding exclusively; a few copies of each title were also printed on vellum. With Walker, he designed three types: a roman, based upon that of Nicolas Jenson, and two Gothics after German models; all were cut and cast by hand.

Woodcut initials and borders were engraved to his own design, and wood-block illustrations were cut from drawings by Edward Burne-Jones and other of his friends.

The Kelmscott Press's major book was its *Chaucer*, finished in 1896, a sumptuous folio whose rich decorations and strong black pages are reminiscent of the German incunabula Morris admired. A table book, meant to be looked at rather than read, it is one of the most influential books in the history of printing--a revolutionary book, despite its anachronisms, which caused a whole generation of printers and designers to be dissatisfied with the books they saw about them and to attempt to improve upon the badly made, weakly designed books that were common in the late Victorian age.

Private presses on the Morris model proliferated in England, on the Continent--especially in Germany and the Scandinavian countries-- and in the United States. The best of these, notably the Doves and Ashendene presses in England and the Bremer and Cranach presses in Germany, published books of great style and strength. There were also poorer imitations, as the Roycroft Press in the United States.

The most influential of the private presses was the Doves Press, established in 1900 by T.J. Cobden-Sanderson and Emery Walker. Walker, who was one of the prime movers in fine printing for over half a century, also played an important role in creating type for the Ashendene and Cranach presses.

214

Cobden-Sanderson was one of Morris' circle at Kelmscott House and had become a bookbinder at the suggestion of Mrs. Morris. The bindings executed at his Doves Bindery are notable for their excellent craftsmanship and their clear, simple design, which often used Art Nouveau motifs (see below). The Doves Press books, which were printed in a type based on Nicolas Jenson's 15th-century roman, were austere in their typography, eschewing all decoration and illustration and relying for their effect on the beauty of their type, spacing, and presswork.

Occasionally a second colour, a splendid red, was used, and superbly drawn initials adorned many of the 50-odd books. A five-volume Doves Bible, issued between 1903 and 1905, is among the monuments of fine bookmaking, as well as one of the most influential modern books, a result of its virility, purity of design, and perfection in craftsmanship.

The third great English private press, the Ashendene, was conducted by C.H. St. John Hornby, a partner in the English booksellers W.H. Smith and Son. Hornby in 1900 met Emery Walker and Sydney Cockerell (Morris' secretary at the Kelmscott Press), who encouraged and instructed him and helped in devising two types for his own use: Subiaco, based upon Sweynheim's and Pannartz' semiroman of the 1460s, and Ptolemy, based upon a late 15th-century German model.

The Ashendene Press books, like those of Morris, were often illustrated with wood engravings, and many had coloured initials.

In Germany Morris' closest counterpart was Rudolf Koch, who gathered around himself at Offenbach, where he taught at the Arts and Crafts School and designed types for the Klingspor foundry, a community of craftsmen who painted, worked in metal, wood, and stone, printed, and wrote. Above all a consummate penman, Koch made the written word the basis of his designs in any medium, whether tapestry or woodcut.

A devout Christian, Koch, like the medieval craftsmen he admired, saw the Gothic style as a supreme manifestation of religious spirit; he was no mere imitator but an artist who freely reinterpreted in his types and books the traditional Fraktur type of Germany. Koch also created a number of modern types, among them sans serifs and romans.

Cobden-Sanderson's influence, however, far exceeded that of Morris in Germany. The most important of the German private presses, the Bremer Presse (1911-39), conducted by Willy Wiegand, like the Doves Press, rejected ornament (except for initials) and relied upon carefully chosen types and painstaking presswork to make its effect. The most cosmopolitan of the German presses was the Cranach, conducted at Weimar by Count Harry Kessler.

It produced editions of the classics and of German and English literature illustrated by artists such as Aristide Maillol, Eric Gill, and Gordon Craig and printed with types by Emery Walker and Edward Johnston on paper made by hand in France. Kessler's books did not attempt to imitate medieval or Renaissance models; they sought to create--using the same methods as the early printers--books modern or, rather, timeless, in spirit.

The most notable figures of the private-press movement in The Netherlands were S.H. de Roos and Jan van Krimpen. De Roos, like Morris a utopian Socialist, was an industrial designer who hoped to create a better society by improving the appearance of ordinary utilitarian objects. His first book, *Kunst en Maatschappij* (1903), was, significantly, a collection of Morris' essays in translation.

De Roos's decorative style became simple and less florid under the influence of Cobden-Sanderson, whose work he greatly admired, although his ideals remained those of the Arts and Crafts Movement. Unlike Morris and Cobden-Sanderson, de Roos was a book designer, designing books for others, rather than a printer--one of the earliest of the new school of typographers, who provided layouts for the publisher or printer, specifying type, format, and overall design.

Increasingly, as technology became more complex and shops more highly specialized and automated, design became more a profession; the typographer, trained in industrial design or graphic arts, succeeded the printer or the publisher in deciding how a book should look. De Roos, who drew a number of typefaces for the Typefoundry Amsterdam, designed books for the Zilverdistel, the Meidoorn, and other private presses, as well as for trade publishers.

Jan van Krimpen used little decoration in his work, which achieved its effect through a classic clarity of style and impeccable printing. His books, for the Enschedé firm for which he worked, for private presses, or for trade publishers, attempted always to interpret the author's meaning as clearly as possible, to reflect it rather than to enhance it. Krimpen also designed a number of typefaces, all of which show his earlier study of calligraphy.

Among them are Lutetia, a modern roman and italic of great distinction; Romulus, a family of text types that includes a sloped roman letter instead of the conventional italic; and Cancellaresca Bastarda, an italic notable for its great number of attractive decorative

216

capitals, ligatures, and other swash (*i.e.,* with strokes ending in flourishes) letters, elegant in appearance.

Another typographer working in the classic mode, Giovanni Mardersteig, spent most of his creative life in Italy, though he was born and trained in Germany. His Officina Bodoni utilized Bodoni's types to print the collected works of D'Annunzio. Mardersteig not only used the handpress for limited editions (usually on handmade Italian papers) that rival 15th-century printing in their beauty of spacing and presswork, but also supervised at the Stamperia Valdònega in Verona long-run editions on high-speed presses, which are likewise remarkable for their craftsmanship. In addition, he designed several typefaces, among them Pacioli, Griffo, Zeno, and Dante.

The Art Nouveau movement was an international style, expressed in the consciously archaic types of Grasset in France; in posters and magazine covers by artist Will Bradley in the United States; and in initials and decorations by Henry van de Velde in Belgium and Germany. Van de Velde, the leading spokesman for the movement as well as one of its most skilled practitioners, in his essay "Déblaiement d'art" (1892) advocated the development of a new art, one that would be both vital and moral, like the great decorative arts of the past, but that would use contemporary modes.

For a reprint of the essay, he designed a series of initials and typographic ornaments that express the characteristics of the style: decoration based upon natural forms; pages whose typography and decoration blend to make overall patterns; and a richness of texture reminiscent of illuminated manuscripts. Van de Velde's most important book was an edition of Nietzsche's *Also sprach Zarathustra*, which he designed for the Insel Verlag and had printed by the Drugulin-Presse of Leipzig and for which he created a series of ornaments printed in gold, as well as endpapers, title page, and binding; a small folio, conceived as an architectonic whole rather than a series of unrelated openings, it is a striking, if dated, volume.

217

25. MASS PRODUCTION AND SOCIETY

Both the quantity and the variety of material goods in industrialized countries have resulted directly from the application of mass production principles. At the same time the environment and circumstances of those employed by, and associated with, the production of material goods have changed.

The benefits that have arisen from the greatly improved productivity made possible by mass production techniques have been shared by employees, investors, and customers. The working environment has greatly changed, however. Similarly the complexities of management have increased substantially, and the investment requirements and risks faced by owners and investors have become much greater.

Before the introduction of mass production techniques, goods were produced by highly skilled craftsmen who often prepared their basic raw materials, carried the product through each of the stages of manufacture, and ended with the finished product. Typically, the craftsman spent several years at apprenticeship learning each aspect of his trade; often he designed and made his own tools. He was identified with his product and his craft, enjoyed a close association with his customers, and had a clear understanding of his contribution and his position in society.

In contrast, the division of labour, the specialization of narrow skills, the detailed engineering specification of how each task is to be carried out, and the assemblage of large numbers of employees in great manufacturing plants have greatly diluted the identification of employees with their productive functions and with their employers. Many surveys in the United States and in the industrialized countries of Europe have shown that workers do not fully understand and appreciate their roles and positions in society. In addition, the division and specialization of labour may lead to such narrowly defined skills and highly repetitive operations, paced by the steady progression of a machine or conveyor line, that tedium and fatigue arise to reduce the sense of satisfaction inherent in productive work.

These physical and psychological factors have been the subject of numerous studies by industrial psychologists and others. Special attention has been paid to work factors which affect the psychological motivation that is a prime determinant of employee productivity. The psychological effects of the repetitive aspects of some mass production tasks have been examined in great detail.

Tasks that are precisely paced by the rhythm of machine operation or conveyor-belt movement appear to be particularly fatiguing. For this reason, efforts are made to structure each job so that the operator can vary his pace by working ahead of the conveyor for a period and then slowing down, and by interspersing work breaks with productive periods.

Some individuals prefer tasks that are sufficiently repetitive and narrowly skilled that they do not require any substantial amount of mental concentration once the function is mastered. Most fatiguing are those repetitive tasks whose pace is out of the operator's control but which also require moderate mental concentration.

With this understanding in mind, work tasks can be structured to produce a minimum of mental and physical fatigue; this planning is an important part of the design of a successful production operation.

The highly repetitive, tightly paced production operations are usually the most easily automated. Thus, as technology advances, that part of the production operation that is most fatiguing, is least satisfying, and takes minimum advantage of the mental and physical flexibility of human effort is replaced by automatic machinery. Not only is productivity improved, but the remaining functions that require human effort can provide a more satisfying experience.

There is also increasing study of the interaction of workers with the tools and machines that assist their efforts. Working together, engineers and physicians are making quantitative biomechanical studies of how the human body functions in performing physical tasks.

These studies, which have led to the improved design of tools and work positions, are part of the broader field of human-factors engineering, which considers the abilities and limitations of people in productive functions and seeks out ways in which machines can be designed to provide the best allocation of function between human effort and machine assistance. These studies are especially important as automated manufacture becomes more common.

The problem of the loss of employee identification with the job has been of special concern. Progressive industrial organizations work to strengthen this identification in many ways, such as by using job rotation and educational programs to diversify employees' experience and to acquaint them with various aspects of the manufacturing process. This can give each employee a concept of the total manufacturing task and the importance of each employee's specific function within that task. Employee suggestion systems provide further opportunity for the individual to have a direct effect on the productive process; the employee is given other opportunities to help structure the manner in which the job is performed.

Thoughtful programs of this type can substantially ameliorate the feeling of anonymity that may otherwise result. Clearly employers must be willing to compromise on the division and specialization of work tasks that technical considerations alone might suggest as desirable. Job content and employee participation must be expanded so that the employee feels significant and retains motivation and identification.

In addition to this increased attention to the structure of the job itself, increases in productivity and resulting increases in wages have reduced working hours and provided employees with opportunities and resources to develop interests outside the workplace. Forward-looking employers, aware of these needs, frequently support these activities through employees' clubs and other means.

END

INDEX   **PAGE:**

Andreas Sofroniou

Andreas Sofroniou

Andreas Sofroniou

BIBLIOGRAPHY - *BASED ON BOOKS PUBLISHED BY ANDREAS SOFRONIOU*

INFORMATION TECHNOLOGY & MANAGEMENT
1. I.T. RISK MANAGEMENT, ISBN: 978-1-4467-5653-9
2. SYSTEMS ENGINEERING, ISBN: 978-1-4477-7553-9
3. BUSINESS INFORMATION SYSTEMS, CONCEPTS AND EXAMPLES, ISBN: 978-1-4092-7338-7
4. A GUIDE TO INFORMATION TECHNOLOGY, ISBN: 978-1-4092-7608-1
5. CHANGE MANAGEMENT IN I.T., ISBN: 978-1-4092-7712-5
6. FRONT-END DESIGN AND DEVELOPMENT FOR SYSTEMS APPLICATIONS, ISBN: 978-1-4092-7588-6
7. I.T RISK MANAGEMENT, ISBN: 978-1-4092-7488-9
8. THE SIMPLIFIED PROCEDURES FOR I.T. PROJECTS DEVELOPMENT, ISBN: 978-1-4092-7562-6
9. THE SIGMA METHODOLOGY FOR RISK MANAGEMENT IN SYSTEMS DEVELOPMENT, ISBN: 978-1-4092-7690-6
10. TRADING ON THE INTERNET IN THE YEAR 2000 AND BEYOND, ISBN: 978-1-4092-7577
11. STRUCTURED SYSTEMS METHODOLOGY, ISBN: 978-1-4477-6610-0
12. INFORMATION TECHNOLOGY LOGICAL ANALYSIS, ISBN: 978-1-4717-1688-1
13. I.T. RISKS LOGICAL ANALYSIS, ISBN: 978-1-4717-1957-8
14. I.T. CHANGES LOGICAL ANALYSIS, ISBN: 978-1-4717-2288-2
15. LOGICAL ANALYSIS OF SYSTEMS, RISKS , CHANGES, ISBN: 978-1-4717-2294-3
16. COMPUTING, A FRÉCIS ON SYSTEMS, SOFTWARE AND HARDWARE, ISBN: 978-1-2910-5102-5
17. MANAGE THAT I.T. PROJECT, ISBN: 978-1-4717-5304-6
18. CHANGE MANAGEMENT, ISBN: 978-1-4457-6114-5
19. MANAGEMENT OF I.T. CHANGES, RISKS, WORKSHOPS, EPISTEMOLOGY, ISBN: 978-1-84753-147-6
20. THE MANAGEMENT OF COMMERCIAL COMPUTING, ISBN: 978-1-4092-7550-3
21. PROGRAMME MANAGEMENT WORKSHOP, ISBN: 978-1-4092-7583-1
22. THE PHILOSOPHICAL CONCEPTS OF MANAGEMENT THROUGH THE AGES, ISBN: 978-1-4092- 7554-1
23. THE MANAGEMENT OF PROJECTS, SYSTEMS, INTERNET, AND RISKS, ISBN: 978-1-4092- 7464-3
24. HOW TO CONSTRUCT YOUR RESUMÊ, ISBN: 978-1-4092-7383-7
25. DEFINE THAT SYSTEM, ISBN: 978-1-291-15094-0
26. INFORMATION TECHNOLOGY WORKSHOP, ISBN: 978-1-291-16440-4
27. CHANGE MANAGEMENT IN SYSTEMS, ISBN: 978-1-4457-1099-0
28. SYSTEMS MANAGEMENT, ISBN: 978-1-4710-4907-1
29. TECHNOLOGY, A STUDY OF MECHANICAL ARTS AND APPLIED SCIENCES, ISBN: 978-1-291-58550-6

EDUCATION & PHILOSOPHY
30. MORAL PHILOSOPHY, FROM SOCRATES TO THE 21ST AEON, ISBN: 978-1-4457-4618-0
31. MORAL PHILOSOPHY, FROM HIPPOCRATES TO THE 21ST AEON, ISBN: 978-1-84753-463-7
32. THERAPEUTIC PHILOSOPHY FOR THE INDIVIDUAL AND THE STATE, ISBN: 978-1-4092-7586-2
33. PHILOSOPHIC COUNSELLING FOR PEOPLE AND THEIR GOVERNMENTS, ISBN: 978-1-4092-7400-1
34. MORAL PHILOSOPHY, THE ETHICAL APPROACH THROUGH THE AGES, ISBN: 978-1-4092-7703-3
35. MORAL PHILOSOPHY, ISBN: 978-1-4478-5037-3

Andreas Sofroniou

36. PSYCHOANALYSIS, POETRY, ISBN: 978-1-4467-2741-6
37. PLATO'S EPISTEMOLOGY, ISBN: 978-1-4716-6584-4
38. ARISTOTLE'S AETIOLOGY, ISBN: 978-1-4716-7861-5
39. MARXISM, SOCIALISM & COMMUNISM, ISBN: 978-1-4716-8236-0
40. MACHIAVELLI'S POLITICS & RELEVANT PHILOSOPHICAL CONCEPTS, ISBN: 978-1-4716-8629-0
41. BRITISH PHILOSOPHERS, 16TH TO 18TH CENTURY, ISBN: 978-1-4717-1072-8
42. ROUSSEAU ON WILL AND MORALITY, ISBN: 978-1-4717-1070-4
43. HEGEL ON IDEALISM, KNOWLEDGE & REALITY, ISBN: 978-1-4717-0954-8
44. PHILOLOGY, CONCEPTS OF EUROPEAN LITERATURE, ISBN: 978-1-291-49148-7
45. THREE MILLENNIA OF HELLENIC PHILOLOGY, ISBN: 978-1-291-49799-1
46. CYPRUS, PERMANENT DEPRIVATION OF FREEDOM, ISBN: 978-1-291-50833-8
47. SOCIOLOGY, CONCEPTS OF GROUP BEHAVIOUR, ISBN: 978-1-291-51888-7
48. SOCIAL SCIENCES, CONCEPTS OF BRANCHES AND RELATIONSHIPS ISBN: 978-1-291-52321-8
49. CONCEPTS OF SOCIAL SCIENTISTS AND GREAT THINKERS, ISBN: 978-1-291-53786-4

MEDICINE & PSYCHOLOGY
50. MEDICAL ETHICS THROUGH THE AGES, ISBN: 978-1-4092-7468-1
51. MEDICAL ETHICS, FROM HIPPOCRATES TO THE 21ST CENTURY ISBN: 978-1-4457-1203-1
52. THE MISINTERPRETATION OF SIGMUND FREUD, ISBN: 978-1-4467-1659-5
53. JUNG'S PSYCHOTHERAPY: THE PSYCHOLOGICAL & MYTHOLOGICAL METHODS, ISBN: 978-1-4477-4740-6
54. FREUDIAN ANALYSIS & JUNGIAN SYNTHESIS, ISBN: 978-1-4477-5996-6
55. PSYCHOLOGY FROM CONCEPTION TO SENILITY, ISBN: 978-1-4092-7218-2
56. PSYCHOTHERAPY, CONCEPTS OF TREATMENT, ISBN: 978-1-291-50178-0
57. PSYCHOLOGY, CONCEPTS OF BEHAVIOUR, ISBN: 978-1-291-47573-9
58. PSYCHOLOGY OF CHILD CULTURE, ISBN: 978-1-4092-7619-7
59. JOYFUL PARENTING, ISBN: 0 9527956 1 2
60. THE GUIDE TO A JOYFUL PARENTING, ISBN: 0 952 7956 1 2
61. PHILOSOPHY FOR HUMAN BEHAVIOUR, ISBN: 978-1-291-12707-2
62. SEX, AN EXPLORATION OF SEXUALITY, EROS AND LOVE, ISBN: 978-1-291-

FICTION AND POETRY
63. THE TOWERING MISFEASANCE, ISBN: 978-1-4241-3652-0
64. DANCES IN THE MOUNTAINS – THE BEAUTY AND BRUTALITY, ISBN: 978-1-4092-7674-6
65. YUSUF'S ODYSSEY, ISBN: 978-1-291-33902-4
66. WILD AND FREE, ISBN: 978-1-4452-0747-6
67. HATCHED FREE, ISBN: 978-1-291-37668-5
68. THROUGH PRICKLY SHRUBS, ISBN: 978-1-4092-7439-1
69. BLOOMIN' SLUMS, ISBN: 978-1-291-37662-3
70. SPEEDBALL, ISBN: 978-1-4092-0521-0
71. SPIRALLING ADVERSARIES, ISBN: 978-1-291-35449-2
72. EXULTATION, ISBN: 978-1-4092-7483-4
73. FREAKY LANDS, ISBN: 978-1-4092-7603-6
74. LITTLE HUT BY THE SEA, ISBN: 978-1-4478-4066-4
75. THE SAME RIVER TWICE, ISBN: 978-1-4457-1576-6
76. THE CANE HILL EFFECT, ISBN: 978-1-4452-7636-6
77. WINDS OF CHANGE, ISBN: 978-1-4452-4036-7
78. A TOWN CALLED MORPHOU, ISBN: 978-1-4092-7611-1
79. EXPERIENCE MY BEFRIENDED IDEAL, ISBN: 978-1-4092-7463-6
80. MAN AND HIS MULE, ISBN: 978-1-291-27090-7

www.ingramcontent.com/pod-product-compliance
Lightning Source LLC
Chambersburg PA
CBHW060834170526
45158CB00001B/161